Gravity Flow Water Supply

Conception, design and sizing for Cooperation projects

First English Edition.
September 2010.

Santiago Arnalich

Gravity Flow Water Supply

Conception, design, and sizing for Cooperation projects

First English Edition.
September 2010.

ISBN: 978-84-614-3277-6

© Santiago Arnalich Castañeda

If you want to reproduce part of the contents of this book, contact us at the following email: publicaciones@arnalich.com.

Cover photo: Marking out the pipeline route in Kabul, Afghanistan.

Translation: Oliver Style
Revision: Amelia Jiménez

Errata at: www.arnalich.com/dwnl/xligraven.doc

arnalich

water and habitat

For Federica and Kelba

"As calming the thirst for justice was a complicated endeavour,
they decided to begin by calming the thirst for water."

Index

1. Introduction

1. 1 ABOUT THIS BOOK

This book intends to provide you with the tools needed to complete a successful gravity flow water project in a short amount of time. You may well already be dealing with a real life project, but without the time to do intensive study to get up to scratch. The book is meant to be:

99 % fat free. With no meticulous explications and never-ending demonstrations. Only what you really need is included.

Simple. One of the most common causes of failure is that complexity and excessive rigor become very intimidating, and things get left half-done

Chronological. It more or less follows the logical order in which you´d undertake a project.

Practical. With calculation examples. For a generous step by step exercise collection see: *"How to design a Gravity Flow Water System. Through worked exercises"*, a book from the same author.

Self contained. It´s assumed you´re in a remote area with no easy access to information, so all the essential information is included. Nevertheless, links to additional sources of information are provided.

Even though canals and aqueducts also work with gravity, they´re not dealt with in this book, so as to make things more manageable.

1. 2 WHAT IS GRAVITY FLOW WATER SUPPLY?

This refers to systems in which the water falls due to its own weight, from a source above, to end-users below. The energy used in the process is the potential energy the water has due to its elevation.

Fig. 1.2a. Digging the trench for the pipeline, Qoli Abchakan, Afghanistan.

The main advantages of this kind of system are the following:

1. There are no expenses incurred through pumping.
2. The maintenance is minimal as there are hardly any moving parts.
3. The pressure in the system is controlled more easily.
4. The systems are robust and reliable.

They are highly suitable to cooperation development projects, as they can supply water to large numbers of people at an accessible cost to local communities.

Mixed projects

Pumped systems tend to be designed to distribute water with gravity flow from a specific point onwards in the system. For example, this system in Somalia pumps water from a borehole to a raised tank; from there, the water is distributed using gravity:

Fig. 1.2b. Mixed system on a nomadic trail. Awr Culus, Somalia.

A typical set up consists of pumping water from a river, lake, reservoir, borehole or well, to a raised tank, from which gravity does the rest. As the water de-pressurizes when it leaves the pipe and enters the tank, the pump has no effect on the water pressure from there onwards. This means the system can be divided into a pumped section, and gravity flow section.

1. 3   TYPES OF DISTRIBUTION SYSTEMS

There are basically two kinds, with distinct characteristics and behaviour:

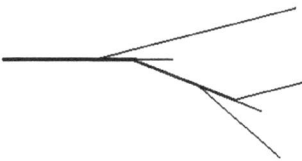

a. **Branched**. These are systems which don´t close loops, and in which the water travels in only one direction. Their main advantage is that they are cheap, quick to build and easy to design. Their disadvantages lie in the fact that a burst pipe cuts off supply downstream; water quality is affected due to stagnation; they are not easy to enlarge at a later date; and the demand at each point has to be worked out very precisely. They are fairly intolerant to design and calculation errors, and risky without reliable data.

b. **Looped**. Loops are closed, allowing the water to travel in any direction. This makes them more difficult to design, but they are more tolerant to design mistakes, more resistant to breakages and with fewer stagnation problems.

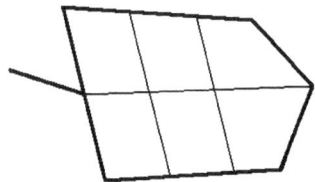

1. 4 WATER AND ENERGY

The entire calculation process of a system is geared towards controlling the amount of energy in the water at any given point, as you need energy to move it from one point to another. In gravity flow projects, the gravitational energy, or **potential energy,** is what moves the water, which is why it´s fundamental to understand what contributes to the amount of energy water has:

The **velocity**. Water which moves has energy due to its movement, kinetic energy. To stop, it requires this energy to be dissipated.

The **height**. If an object is a few meters off the ground, it has more potential energy than one which is at ground level. It could fall, converting the potential energy into velocity, or kinetic energy. If it has velocity and meets up an uphill slope, it will move up the slope until all its kinetic energy is transformed into potential energy.

The **weight of a column of water**. In a mass of water, each molecule supports the weight of the ones above it. This weight will increase its energy. If the water is in a U-shaped tube, this weight will make it rise up the other side of the U until it´s at the same height.

All of this is summed up in the Bernoulli equation: $\quad H = Z + \dfrac{P}{\gamma} + \dfrac{V^2}{2g}$

H is the total energy in the system, expressed in *meters of a column of water*, and each of the three terms follow exactly the parameters mentioned above:

$+\dfrac{V^2}{2g}$ Velocity ; $+\dfrac{P}{\gamma}$ Weight of Column Z ;   Height

In gravity flow water systems, the velocity is very low and the velocity component can be ignored. The height can be measured.

The pressure is the force (weight in this case) by the surface area on which it is being applied. The weight depends on the volume of water. As an increase in the surface area increases the weight and the surface area on which it is applied in the same proportion, the height of the column of water is the only element which changes the pressure. This means the pressure can be expressed in terms of meters of a column of water (*mca*), which is more comfortable and intuitive. At a depth of one meter, the pressure is one meter. At ten meters, the pressure is 10 meters.

The Bernoulli equation becomes less intimidating and simplifies to this:

H = Height + Depth

When water is sitting still in a container, it´s surface is horizontal and it has the same amount of energy at all points, a constant:

H = Height + Depth = constant

This is independent of the shape of the container holding the water.

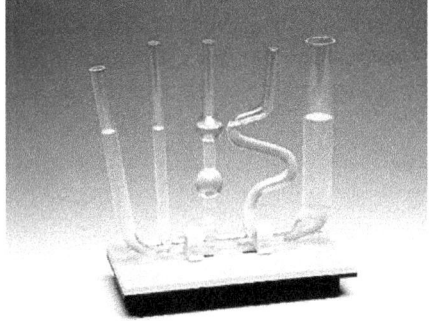

Hydraulic gradient J

Now imagine a pipe on a slope. The energy at point *A* is due to its height, 30m. At point *C*, due to the column of water above it, of 30m. At point *B*, the column is 20 meters and the depth 10 meters, which means the energy is the same as a column of water of 30 meters.

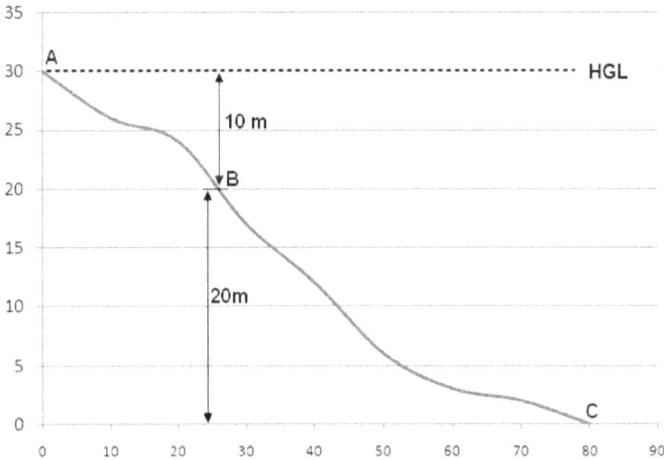

At all points the energy is the same, the equivalent of a 30m column of water, and the joining of these points forms a line we call the **hydraulic grade line (HGL).**

Water in movement

When water begins to move inside a pipe, it rubs against the walls of the pipe and loses energy in the form of heat. The quantity of energy lost depends on its velocity and the roughness of the pipe. This is expressed as X meters of a column of water lost over each kilometre of pipe.

This parameter is called the **head loss, J**. If, for example, the water has travelled 1000m and has lost 10m of a column of water, J=10m/km, and the line of the hydraulic gradient inclines downwards:

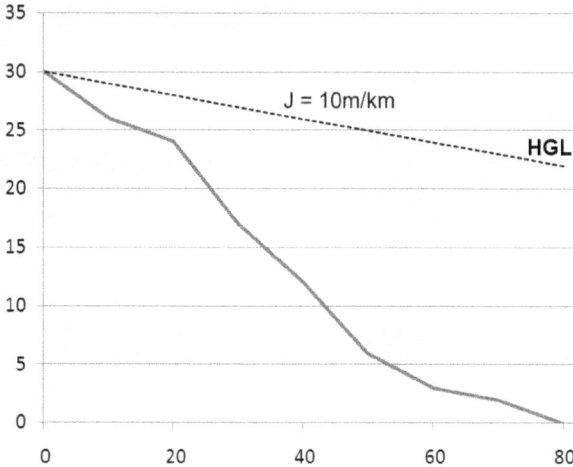

And this is where the fundamental rule of gravity flow water projects comes into play:

To maintain pressure in the system, the hydraulic gradient must always be 10 meters above ground at any point[1].

Using smaller pipe diameters increases the friction loss and inclines the HGL further. Using larger pipe diameters means the HGL rises to become more horizontal.

Units

Before going any further, memorize this relationship. You´ll need it to relate the results of your calculations with the pipes available on the market:

$$10 \text{ meters of a column of water (mca)} = 1 \text{ kg/cm}^2 = 1 \text{ bar}$$

_____

[1] There are exceptions. A very obvious one is at the start of the pipe, where the terrain has still not dropped 10m.

1. 5   THE HANG GLIDER ANALOGY

Try imagining the design of a gravity flow pipeline as a flight in a hang glider. Taking off from a raised position where the source is, you have to control the fall so as to reach your destination at the right height.

Fig. 1.5. Taking off, the hang glider analogy.

The fall rate is the head loss and the final height is the design pressure at the final point of use. If this has been established as 1.5 bar, you need to *land* 15m above the ground, making sure you fly at least 10m above ground on the way, avoiding all the obstacles:

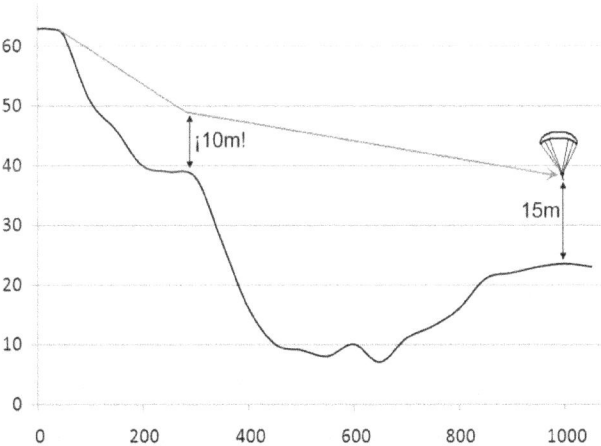

1. 6   AN INITIATION FLIGHT

Calculation example:

A distribution system is planned, of PVC pipe, with 4 l/s at 2 bar of pressure, over a distance of 1350m. The topographic profile is shown below:

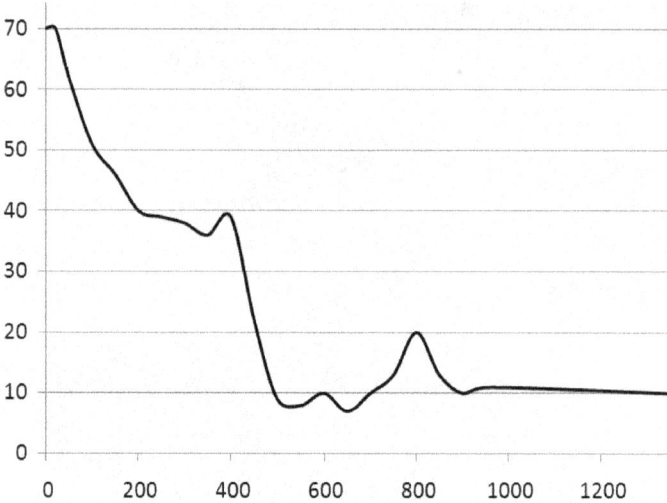

The inlet is at 70m, and the *landing* will be at an elevation of 10m, together with 20 meters of pressure: 30m. There are 2 points to get over:

a.   The first point has an elevation of 40m at a distance of 400m. To fly over it with 10m to spare, the maximum fall you can afford is:

   70m – (40m + 10m) / 0.4km = 50m/km

In the tables of Appendix B, you need to find a pipe which will give you a friction loss of 50 m/km or less, at a flow of 4 l/s. At the start of the appendix you´ll see an explanation of how to use the tables. Assume the water is clean (k=0.01) and that you are using PN10 pipe.

Seeing as there isn´t an infinite range of pipes available choose the closest one, Ø63mm, which will give you a drop of 45m/km.

PVC 63 - ID 57mm- PN 10		
J (m/km)	Q (l/s)	v m/s)
30,00	3,236	1,27
45,00	4,049	1,59
60,00	4,744	1,86

The height at which you'll fly over point *a* is:

70m – (0.4km * 45m/km) = 52m

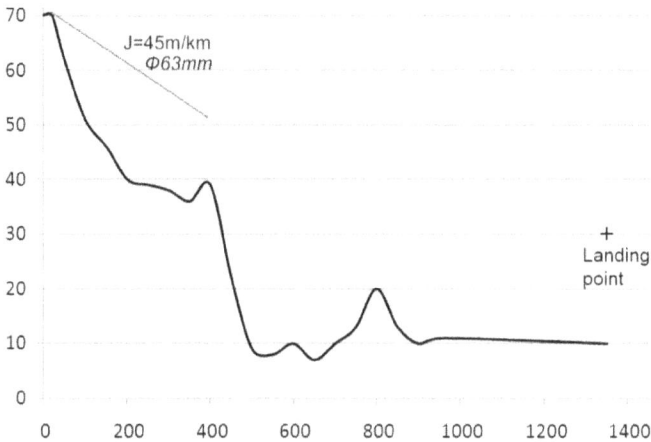

b.   The second point, at 800m and 20m of elevation, isn't really a major obstacle, as its height with 10m added is the same as the landing point. We can aim straight for the landing point. The horizontal distance is:

1350m - 400m = 950m

The maximum drop will be:

52m – (10m + 20m) / 0.95km = 23.16m/km

We won't be able to use the same pipe all the way there. If you choose 90mm, the head loss will be only 8m/km, and the water will arrive with too much pressure. If we use 63mm with a head loss of 45m/km, we'll lose too much height too quickly and the water will arrive with insufficient pressure.

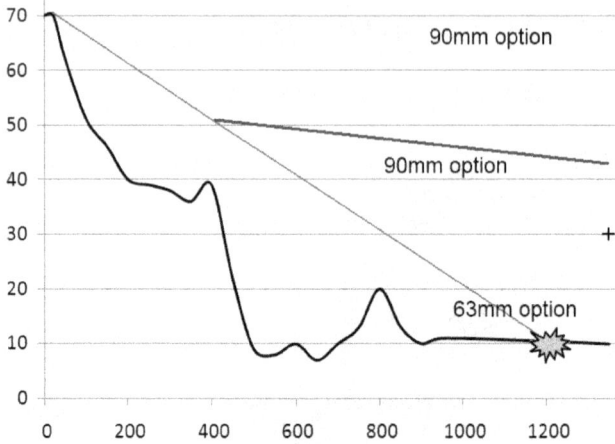

For the water to arrive with the right amount of pressure, a second section, or *reach*, is put in place by combining pipe sizes. The most logical would be to carry on with 63mm pipe up until point *X*, and then continue with 90mm. To work out where this point is, you need to use the following equation:

Head loss for *X* km of pipe *A* + head loss for the remaining distance of pipe *B* = maximum possible drop

$$X * 45 \text{ m/km} + ((0.950\text{km} - X) * 8\text{m/km}) = 52\text{m} - (10\text{m} + 20\text{m})$$

$$45X + 7.6 - 8X = 22 \;\rightarrow\; 37X = 14.4 \;\rightarrow\; X = 0.389\text{km}$$

In the second reach, we can use 389m of 63mm pipe, and 950-389=561m of 90mm pipe.

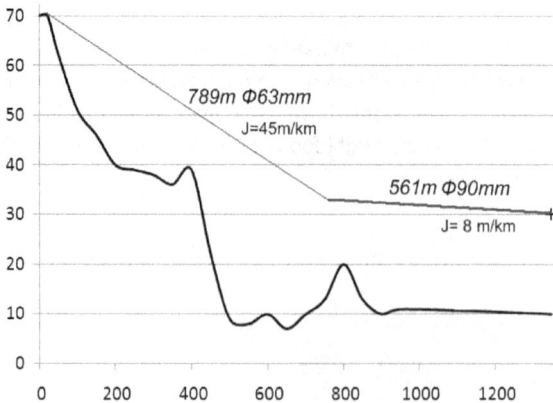

Congratulations! Now you´ve got the basic tool for designing gravity flow water projects. From now on, it´s just a matter of adding various different elements to make sure you use it properly. If your flight has been a little turbulent or you missed a bit along the way, go back over this chapter, as it's fundamental to your understanding of the rest of the book. You´ll find more examples in Chapter 5.

1. 7 WORKING OUT THE UNITS AND AVOIDING MISTAKES

For a coherent design you´ll need to make a number of very simple calculations by hand. They may be simple, but it´s easy to make mistakes, and they can be as deceptive as double negatives or trying to work out the number of days between 2 dates in the calendar.

If you´re disciplined with the units, you´ll spot most of these mistakes before you have a nervous breakdown. Have a look at this example, of two different ways of converting units of m^3/h to l/s:

$$14 \ m^3/h = 14 \ m^3/h * 1m^3/1000 \ l * 3600s/1h = 14*3600/1000 \ m^3*m^3*s/h*l*h$$

$$= 50.4 \ s*m^6/ \ h^2*s$$

¿¿ $s*m^6/ \ h^2*l$?! If you´re like me, then this unit makes no sense, and something´s gone wrong...

$$14 \ m^3/h = 14 \ m^3/h * 1000 \ l/1m^3 * 1h/3600s = 14*1000/3600 \ m^3*l*h/h*m^3*s$$

$$= 3.88 \ l/s$$

Note that the two results are a little different.

NOTE: Multiplying by 1h/3600s is the same as multiplying by 1/1, as 1 hour and 3600 seconds are the same. If it makes things easier, think of it as "in every hour there are 3600 seconds." The result is a change of units.

2. Population and demand

2. 1 ENTHUSIASM AND CONFLICT

A population is little more than a group of people living together, each with their criteria, weaknesses, motivations and worries. Although gravity flow water systems are relatively simple to organize, they require social cohesion and organizational capacity. It´s essential to evaluate the local population´s work capacity and enthusiasm, as they´re key factors in the success or failure of a project.

Doing this however, is not that simple. If you observe existing social structures (this and that organization, women´s committee...) and the collective efforts made in the past, you can get an idea.

On the other hand, a project can generate powerful emotions and water is often a source of conflict. To ignore the needs of a collective group can lead to resentments and even sabotage of completed systems. Imagine, for example, that a water project uses almost all the available water, leaving farmers downstream with nothing.

2. 2 DESIGN PERIOD

No structure will last forever, and an important decision is that of the design period. In other words, how long the system will be in service for. This decision is important, as it will determine how many people will be served.
If the system is projected taking into account the current population, it will be outgrown before it´s even been built. This is why you need to design a system taking into account how big the population will be at the end of the design period. Another

important factor, which you´ll see more of later, is the economic calculations. An investment of € 50,000 over 5 years is not the same as over 50 years.

Normally, 30 years is taken as the design period, despite the fact that the minimum projected life of PVC, for example, is 50 years. Projecting beyond 30 years entails much greater uncertainties, and a bigger initial investment. You don´t believe me? Have a look at this:

> *"I think there is a market for around 5 computers in the whole world."*
>
> (Thomas Watson, President of IBM, 1943)

Or this, made 30 years ago:

> *"There´s no reason why a normal person would want a computer in their home."*
>
> (Ken Olsen, pioneer in the development of computers, 1977)

2. 3 FUTURE POPULATION

To work out what the population will be at the end of a design period consists of projecting the population according to some kind of parameter, such as population density or growth rate.

Projection based in growth rate

There are various formulas. The one which is most widely applicable and which probably works best is the geometric. Here you have a comparative result:

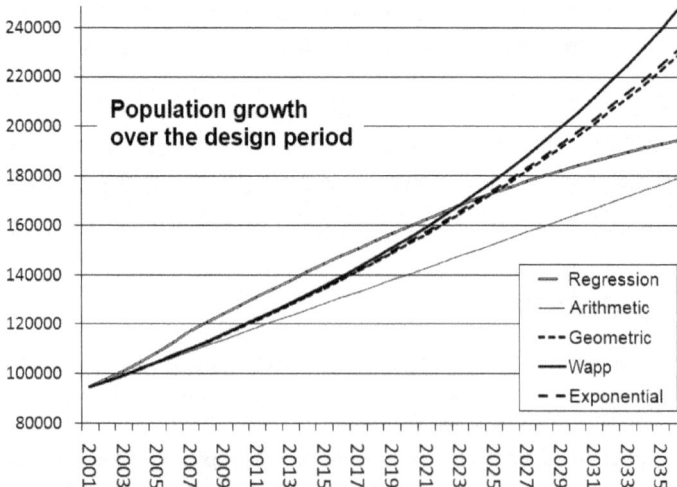

Population growth over the design period

If there is a census, population data and growth rates can be easily worked out. Otherwise, you´ll have to do a survey. Unfortunately, most populations have a really distorted idea of their own numbers, and in cooperation work this kind of data is often exaggerated to gain support and funding. Avoid the temptation of using country-wide rates, as local populations can grow at varying rates.

Three projection formulas commonly used are:

Arithmetical: $P_f = P_o \left(1+\dfrac{i*t}{100}\right)$

Geometrical: $P_f = P_o \left(1 + \dfrac{i}{100}\right)^t$

Exponential: $P_f = P_o * e^{\left(\frac{i*t}{100}\right)}$

P_f , future population
P_o , current population
i , growth rate in %
t , time in years
e , number e, (e=2.71828...)

Calculation example:

The current population of Mwata is 3780 people. 5 years ago, the population census counted 2703. What will the population be for a design period of 25 years?:

From the census, the growth rate can be estimated:

3780 people – 2703 people / 5 years*2703 people = 0.0796

The growth rate is around 8%.

Applying the geometrical projection formula:

$Pf = P_o (1+ i/100)^t = 3780 (1+ 8/100)^{25} = 25,887$ inhabitants

Applying the arithmetical projection formula:

$Pf = P_o (1+ i*t/100) = 3780 (1+ 8*25/100) = 11,340$ inhabitants

Applying the exponential projection formula:

$Pf = P_o * e^{(i*t/100)} = 3780\ e^{(8*25/100)} = 29,731$ inhabitants

---

Projection based on density

The formula projections can overestimate future population and makes things cumbersome, especially in areas which have been recently populated or where there is intense immigration.

An alternative focus is to assume that cultural characteristics and the economic situation of the population will impose a limit on population density. The population will tend to reach a density limit, at which point it will begin to feel overcrowded. As the population moves closer to the density limit, the growth rate drops until it reaches 0. It´s easier to look at the following example:

Calculation example:

The block UV39 is a poor area of Santa Cruz recently populated, on what used to be arable lands between the airport and the industrial zone. The average family has 6 people. What design population would you use if the system is planned for 30 years, taking into account the following data from the last census?

A99	A104	A109	A114
A100	A105	A110	A115
A101	A106	A111	A116
A102	A107	A112	A117
A103	A108	A113	A118

Block 39 plan

Plots	Families	Growth rate
A99	36	1.92
A100	35	2.85
A101	33	2.8
A102	36	0.96
A103	34	1.91
A104	36	1.5
A105	35	1.35
A106	32	2.58
A107	36	1.7
A108	36	0.46
A109	10	10.1
A110	9	17.49
A111	29	3.69
A112	35	1.35
A113	36	1.2
A114	12	11.32
A115	3	11.83
A116	4	15.44
A117	9	10.6
A118	8	15.2

Source: 2005 Census

Note how, in general, the growth rate is lower the more families there are on each plot. This tendency means you can work with the theory that there is a value for the density limit which the population considers acceptable. You need to work out what that limit might be. It´s a good idea to talk with the local population and ask them if they feel tightly packed or not, so as to avoid false limits which are too low. We could take the value of 36 families per plot, or slightly higher:

20 plots * 36 families/plot * 6 people / family = 4,320 people.

If you´d used the geometrical formula in this case, taking an average growth rate of 5.8%, you would have ended up with 16,412 people- almost four times as much.

Always pay close attention to the dynamics of a population. In this case, the least populated plots are concentrated in one area:

A99	A104	A109	A114
A100	A105	A110	A115
A101	A106	A111	A116
A102	A107	A112	A117
A103	A108	A113	A118

By investigating the causes, it may simply be a question of access to services, but it may also be due to periodic flooding in the area, or even mines. Delivering services stimulates growth, and close attention should be paid to ensure that new areas are not dangerous.

A further complication to achieve greater precision is to do a lineal regression to work out the density limit. Exercise 22, reference 2, in the bibliography shows you how.

2. 4 DAILY BASE DEMAND

This is the quantity of water the population will consume, for all kinds of uses: cooking, washing, drinking, work activities...There´s no quick fix to working out the demand of a specific population. As a guide, below are some **minimum** figures with which to work:

Minimum daily demand (l/un.)	
Urban inhabitant	50
Rural inhabitant	30
Student	5
Outpatient	5
Inpatient	60
Ablution	2
Camel (once a week)	250
Goat and sheep	5
Cow	20
Horses, mules, and donkeys	20

You want to know how much water a lawyer's firm consumes? It's unlikely you'll need to know this, but if so, you can find out the estimated demands of unusual suspects in table 4.1 of reference 19, taking into account that these are demand figures for the USA. The complete book can be found here:

http://www.haestad.com/library/books/awdm/online/wwhelp/wwhimpl/js/html/wwhelp.htm

An area of demand you need to look closely at to avoid unpleasant surprises is that of small vegetable gardens. Most species consume around 5mm/m^2 daily. Seeing as 1 mm/m^2 is the same as 1 l/m^2, a small garden of only 20m^2 consumes around 100 litres a day. If vegetable gardens are common, they can add up to a considerable proportion of the total demand.

In practice, the idea is to supply the maximum quantity of water such that:

- environmental health problems are not created (stagnant pools, over exploitation of the source)
- people are prepared to pay for the service
- the cost is appropriate to the local ability to pay.

To sum up, if I have 2 goats, 3 rural inhabitants and 1 donkey, the total daily demand is:

2 goats x 5 l/goats*day	= 10 l/day
3 people x 30 l/ people*day	= 90 l/day
1 donkey x 20 l/donkey*day	= 20 l/day
	----------------------
	120 litres/day

This figure is expressed in litres per second as the **base demand**, protagonist of the rest of the chapter:

120 litres/day * 1day/24h * 1h/3600s = 0.00139 l/s

2. 5   TEMPORAL VARIATIONS IN DEMAND

How much water is consumed on a daily basis is as important as *when* it is consumed. If at 9:00am twice as much water is consumed as the daily average, the carrying capacity of the system will have to be doubled to cover the peak demand period.

Daily variations

Most populations follow a similar dynamic. Night time demand is minimal, with demand peaking in the morning. People are showering, collecting water for cooking or washing, and can often consume 45-65% of their daily use in only a few hours. Around the middle of the afternoon there's another smaller peak of roughly 20-30%.

This was the daily demand pattern of a low-income urban population in Santa Cruz, Bolivia:

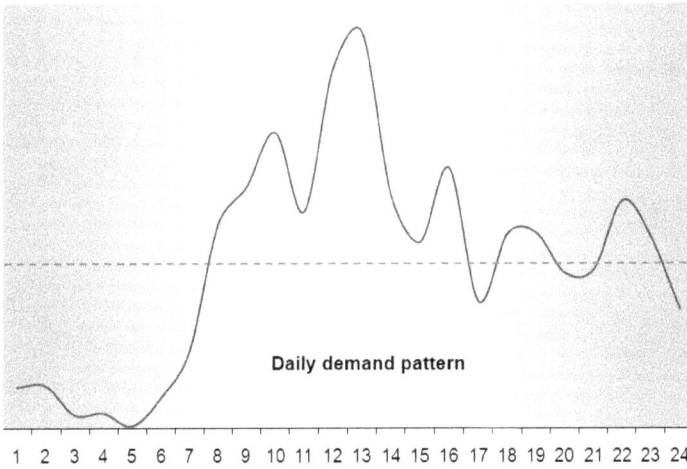

Daily demand pattern

1 2 3 4 5 6 7 8 9 10 11 12 13 14 15 16 17 18 19 20 21 22 23 24

Measuring the demand variations of a population can be time consuming and complicated, and it's not always possible. A large number of consumers need to be measured at the same time to establish the overall demand pattern. Look at the differences between the demand patterns of the 30 consumers in Bolivia used to construct the previous demand pattern:

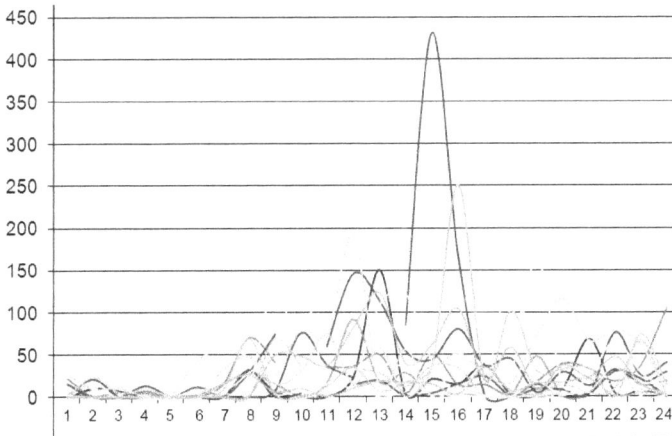

There's often not even a working system in place to measure from!

Luckily, most populations, whatever their demand pattern, usually have a peak demand of around 2.5 times more than the average. To take into account the daily variations and make sure the system can deal with the daily peak, multiply your average demand by 2.5:

$$0.00139 \text{ l/s} * 2.5 = 0.0035 \text{ l/s}$$

Weekly variations

Among most populations there are no big variations, but keep an eye out for cultural changes, parties, markets, fairs etc. They can leave their mark on the daily demand at specific points in the week. Among the same population in Bolivia, the demand tends to reduce as the week goes on:

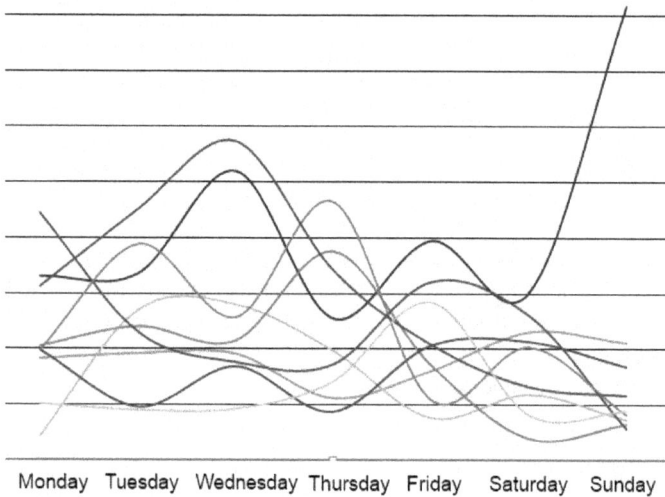

Monday Tuesday Wednesday Thursday Friday Saturday Sunday

If the daily pattern was difficult to measure, the weekly one (which requires effort and logistics during an entire week), is really tricky. Unless you have a clear indication of weekly changes, you can assume the demand doesn´t change much during the week. The simplest solution may lie in measuring the amount of water that comes out of a tank during a week.

Monthly variations

In sharp contrast, these differences tend to be measured very closely, as they are the basis for the billing of the water supply. Any nearby system which bills their customers will be able to give you very precise information. The variations are important, most of all during the different seasons. Look at how the demand falls with lower temperatures in the southern winter during the years 2002, 2003, and 2004:

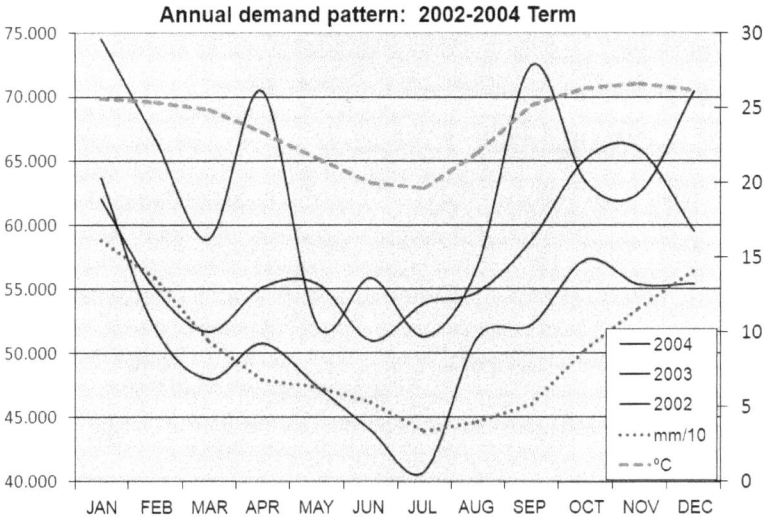

Annual demand pattern: 2002-2004 Term

To work out how to increase the base demand, there are 2 paths you can take:

a. If measurements were taken on a specific day for the daily demand pattern, compare the month of maximum demand with the month you made the measurement in. For example, if you measured in July and the bills show a total demand of 600 m³, and the maximum demand over the year was 840 m³ in October:

$$840m^3/600m^3 = 1.4$$

The adjusted base demand would be: 0.0035 l/s * 1.4 = 0.0049 l/s

b. If you don´t know which month the measurements were taken and you´re using a generic multiplier like 2.5, work out the average of all the months and compare it with the month of maximum demand, as above.

2. 6 UNACCOUNTED FOR WATER

This category is a box of surprises which includes leaks, unauthorised connections, garden watering, water which is spilt when containers are filled etc. In a new system, it´s around 20%:

0.0049 l/s * 1.2 = 0.00588 l/s

If you´re fixing an old system, you can get an idea by looking at the night time demand. At night hardly anyone uses water, and the pressure rises. If between 2am and 5am, the demand is more than 3% of the daily total, you can be pretty sure there are some major leaks.

2. 7 FIRE FLOW DEMAND

This relates to the water available in case of fire, and is done by making sure that at all times there is:

- A **fire reserve**, an amount of water stored exclusively for that purpose. The standards vary from one country to the next, but normally it´s the equivalent to the fire flow during 2 hours.

- A **fire flow**, which is worked out depending on the population type and how big it is.

In practice, the requirements are so huge that the fire demand is usually the one which ends up determining the size of the system, as it´s often many times more than the peak demand from the population. In cooperation projects I´ve seen 2 different approaches: either fire flow demand is ignored altogether, or there is strict adherence to the western standard.

To ignore the need for protection from fires is so reckless it requires no further comment. However, applying western standards or that of the country you´re working in often blows things out of proportion. Does it make sense to ensure a fire flow of 32 litres per second in a community which only has buckets to fight a fire with?

I think there´s a middle ground. It´s not easy to work out what that is. Either way, it´s a good idea to talk to the local fire brigade and see what they think, and by fire brigade, I mean the people who are going to be putting out a fire. This doesn´t mean they necessarily wear a helmet and run around in their fireman´s outfit.

More than a specific flow, the important thing is to make sure there´s a quantity of water stored so that a small fire doesn´t turn into a catastrophe because the tanks are all left dry as a bone.

2. 8 SUMMING UP: PREPARING FOR THE WORST

Once you´ve worked out the future population, you can see that the different average demands of the future population has been multiplied by varying coefficients, to take into account temporal variations and specific demands:

Base demand:			0.00139 l/s
Daily variations:	0.00139 l/s * 2.5	=	0.0035 l/s
Weekly variations:	0.0035 l/s * 1	=	0.0035 l/s
Monthly variations:	0.0035 l/s * 1.4	=	0.0049 l/s
Unaccounted for demand:	0.0049 l/s * 1.2	=	0.00588 l/s
Fire flow demand:	???	=	???

Once all the coefficients have been applied, the demand has increased more than 4 times, from 0.00139 l/s to 0.00588 l/s.

The idea behind this approach is to prepare for the worst situation the system will face: if it works then, it´ll work all the time. In other words, if a pipe is able to transport 10 l/s, it can also transport 2 l/s.

If the system can deal with the worst hour, of the worst day of the week, of the worst month, for the population following 30 years of growth, it´ll work at any other time.

No data

Imagine you´re in a town in the middle of nowhere in East Timor where the inhabitants walk 6km to fetch water every day from a spring. Hummm, where do I get the data from?!

If you can´t find data on the daily, weekly and monthly variations and the unaccounted for demand, an approximation is to multiply the base demand by a number between 3.5 and 4.5:

0.00139 l/s * 4 = 0.0056 l/s

See how it´s very similar to the result we found from the population in Bolivia.

2. 9   SMALL SYSTEMS

Pay close attention to this section, as it´s really important. When a system is small, if you use the average flow as I´ve just described, you´ll end up installing pipes which are too small. Imagine just one tap at the end of a pipe with a flow of 0.2 l/s. If at the end of the day you´ve supplied 50 litres, the average flow is very small:

50l / 24h*3600 s/h = 0.00058 l/s

However, when the tap opens, the pipe has to carry 0.2 l/s...almost 350 times more!

This difference between the average flow and the instantaneous flow becomes smaller and smaller as the number of users increases. The fact that one user opens or shuts a tap becomes less and less important, and the average demand begins to stabilize. The point at which they are the same varies from one place to the other, but you can use the rule of thumb of 250 connections (not people).

This effect takes place in every pipe. Even if a system supplies a lot of people, if in one particular branch line there are only 35 connections, be careful not to install pipes which are too small.

The simplest way of avoiding problems is to establish a **minimum diameter** for the entire system. For demands of over 50 litres per person per day, use 3"; less than 2" won´t be enough. Using minimum diameters is also really useful for protection against fires.

For large systems, there´s no excuse for using pipes smaller than the minimum diameter, as you´re working with a bigger budget and the added expense is minimal. The concentration of more people also demands greater protection from fires.

For smaller systems, if your budget allows, use minimum diameters. If you´re working with a tight budget, as is often the case, design the system for:

1. *All taps open*. This is for systems where queues are expected like refugee camps, emergencies or public tapstands. The design flow is the addition of all taps that can be opened and temporal variations are ignored altogether.

Calculation example:

You want to install a galvanized iron pipe 1km long to supply 3 taps. The topographic survey reveals that the pressure can fall by 4.15m. What diameter of pipe should be installed to make it as cheap as possible?

The standard tap flow is 0.2 l/s, so the total is 3*0.2 l/s = 0.6 l/s.
The maximum head loss is 4.15m/km.

Since we´re dealing with a small system and for economic reasons, we´ll calculate for *all taps open*. From table C we can see that it´s 2" for 0.6 l/s

---

2. *Simultaneity*. If queues are not expected but the number of connections is small the design flow is obtained by multiplying the average demand by a simultaneity coefficient you can get from this graph (Arizmendi 1991):

Multiplier

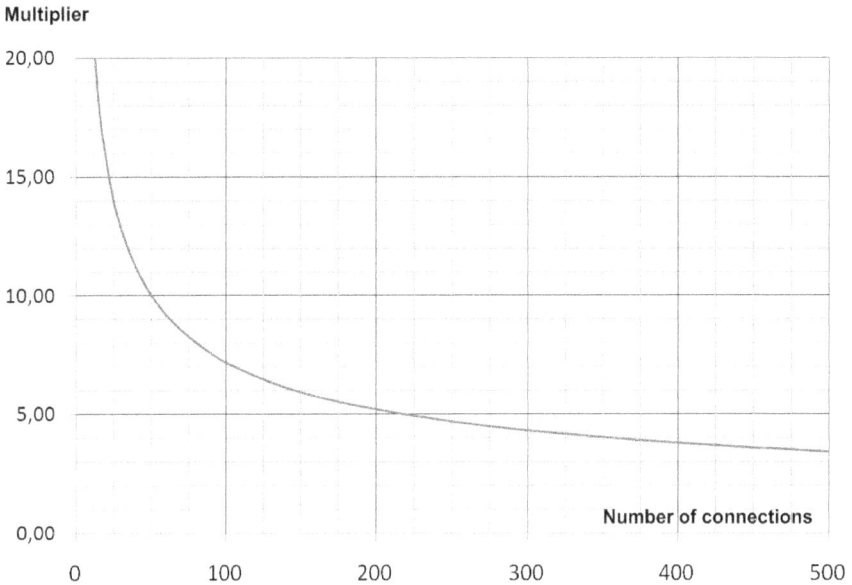

To sum up, choosing the method for calculating the design demand comes down to 3 basic questions, which you can see in the flow diagram below:

Calculation example:

What would be the design flow of a pipe with 100 connections if 86,400 liters are consumed daily from it?

The average demand is 86,400 l/day * 1 day/24h * 1h/3,600s = 1 l/s

For 100 connections, we get a coefficient of 7.2 in the graph. The design flow is:

Q= 1 l/s * 7.2 = 7.2 l/s

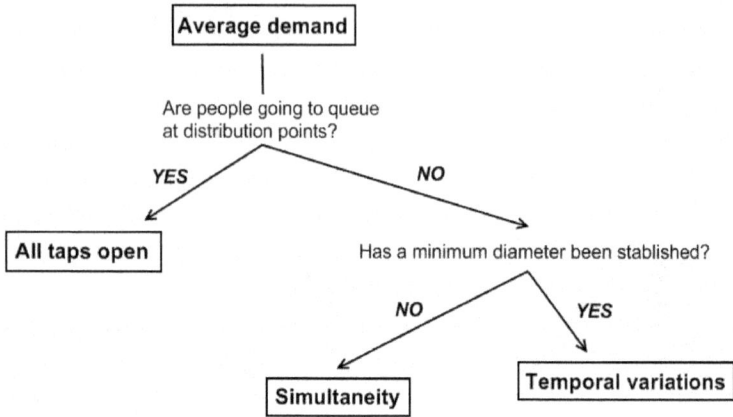

3. Sources

3. 1 TYPES OF SOURCES

The sources most frequently used in gravity fed distribution are small streams and springs. A *source* can also be a reservoir tank in mixed systems.

Springs

A spring is a place where underground water surfaces, usually on the side of a hill or mountain. The water filters slowly down through the subsoil until it reaches an impervious layer, usually rock. If this layer meets ground level, the water flows out forming a spring.

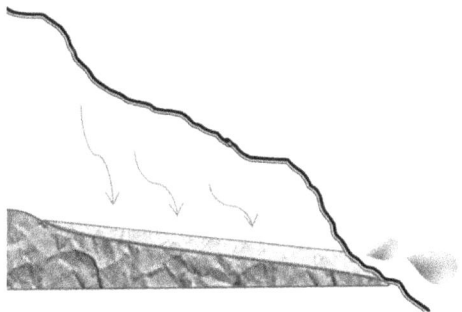

The big advantage of springs is that the water has been purified and filtered on its journey below ground, and doesn´t need treatment. This means illness due to errors in chlorine dosing is avoided, costs are lower and complicated logistics are avoided. One notable exception is springs in fractured rock in which the water has not necessarily been filtered.

Streams

Streams are surface water sources. All surface water should be taken as contaminated, requiring treatment.

Don´t underestimate the potential of tiny streams. This photo shows a stream which supplied 45,000 people in the refugee camp of Mtabila (Tanzania), and survived a drought which dried up all other surrounding sources. However small a stream may be, the quantity of water it carries is relatively large. It´s important to take into account seasonal variations.

To avoid having to treat the water, the intake works can be built to function as a filter. In small rivers and streams, one option is to dig a traditional well some meters from the edge in a flood-protected area. Otherwise, you can divert a part of the flow and pass it through a sand filter before it enters the captation tank.

3.2   SEASONAL VARIATIONS

The chosen source for your project must maintain sufficient flow **throughout the whole year.** This may seem obvious, but it´s the most common cause of failure in cooperation projects.

Fig. 3.2. Dry main source, Mtabila II refugee camp, Tanzania.

Unfortunately, the size of the source is a poor indicator of its seasonal variation. Streams which look huge can dry out completely a few weeks later. Small streams you may have discarded in favour of larger ones often maintain their flow rate.

Perhaps the easiest and safest way is to ask the local population, above all the people who depend on water (farmers, local inhabitants…).

3. 3 FLOW MEASUREMENT

The simplest way of measuring the flow is with a container of a given volume and a watch. However, the flow is often too big and makes this impractical.

V-notch weir

This is one of the most practical ways of measuring flows above 2 or 3 litres per second. It involves making the water flow through a weir in the form of a V:

Fig. 3.5. Stream flow measurement. Dongwe Project, Tanzania.

The height reached by the water (measured with a ruler) is proportional to the flow. To determine the flow use this formula:

$$Q = 533 * C_e \sqrt{2g} * h^{2,5} * \tan(\beta/2)$$

Q, flow in l/s.
C_e, coefficient, dependent on the construction. Normally, 0.64.
g, gravity, 9.81 m/s^2.
h, height of the water in meters.
β, angle of the weir in radians[2].

In normal conditions, g= 9.81 and C_e=0.64, so the equation simplifies to:

$$Q = 1510 * h^{2,5} * \tan(\beta/2)$$

For 60° and 19 cm, it will be:

60° * 0.01744 = 1.046 radians
Q= 1510 * 0.19^{2,5} * tan(1.046/2) =13.7 l/s

---

[2] Obtained by multiplying the degrees by 0.01744. E.g. 60° * 0.01744 = 1.046 radians.

If you don´t want to use tangents of angles in radians, build a 60° weir and use this graph:

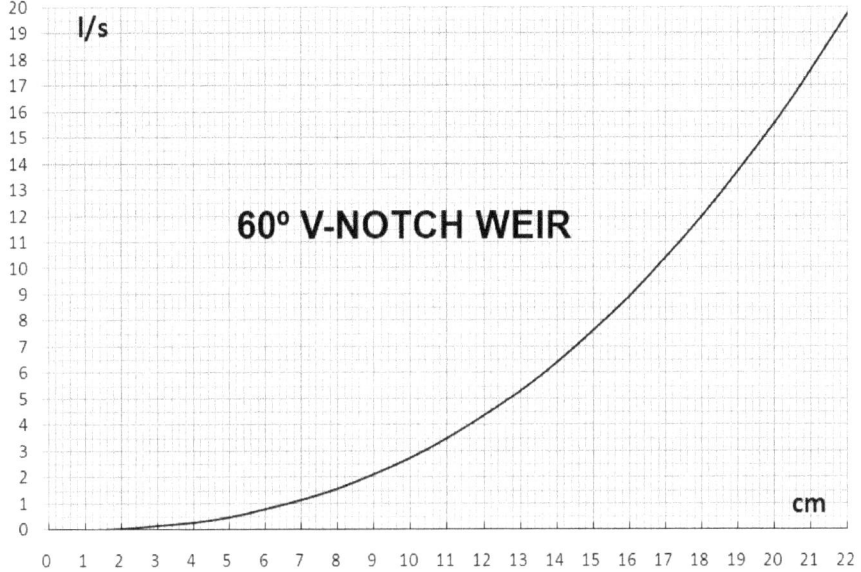

Others

It´s unlikely you´ll need them. The maximum on the graph is sufficient to supply around 35,000 people with 50 l/person. If you really want to measure the flow of a larger stream, you can use the floater method. Throw a floating object into a straight bit of the stream, whose cross section you can calculate. Measure the velocity and apply this:

$$Q = 850 * v * A$$

Q, flow in l/s
v, velocity in m/s
A, cross section m^2

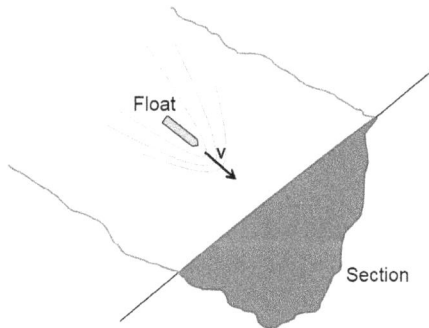

An alternative but unlikely method is that of dissolving salt.

Explotation limits

Most countries have legislation or very clear guidelines on how much you can extract from a source. In the case of refugee camps in Tanzania, for example, you can *only* extract 80% of the flow. For practical purposes, taking 80% of the flow of a stream leaves it almost dry, with potentially disastrous consequences further downstream: loss of vegetation, erosion, and depletion of nearby sources. In this case, the human consequences were far worse, and 80% was the capacity used.

Before projecting the extraction from a source, even with much smaller percentages, make sure you investigate what happens further downstream:

> Is anyone using the water?
> Is it recharging an aquifer?
> Is there vegetation or forest on the banks? What is their value?

People often think that water flowing past is water lost. However, this water plays a key part in the local environment and should be exploited in exchange for tangible and clear benefits.

3. 4 WATER TESTING

To guarantee the safety of the water source and the viability of the project you need to get the water tested. Appendix A contains a summary of the parameters and their maximum values, as recommended by the World Health Organisation.

Collection of samples

It´s important to collect a sufficient quantity, so the laboratory can do the tests. One and a half litres is enough. An empty mineral water bottle is ideal. Avoid using old or metal containers. Either way, consult with the laboratory as to the quantity requirements and conditions of transport.

> **Tap:** Remove any accessories that may have been added to the tap and clean the end with a cloth. Let the water run for a few minutes and take the sample without letting the container touch the tap.

> **Lake:** Take the sample 30cm from the surface, to avoid collecting pollutants which either float or are sedimented. Make sure no part of your body makes contact with the water!

> **River:** If possible, take the sample going against the flow, 30cm from the surface, avoiding stagnant areas or anywhere near the banks.

Biological testing

In my opinion these tests don´t influence the choice of source all that much. They often give false results and the amount of bacteria from a source can vary greatly over time. You run the risk of taking a source as safe, when in fact it may become contaminated again every time it rains.

If the water is surface (rivers, lakes, streams…) it will almost certainly be contaminated at some point. Moving waters are less dangerous than stagnant ones. If this is the case, you´ll need to build a filtering intake, or establish a chlorination system.

If the water has a residual chlorine concentration of 0.2 ppm, a turbidity of less than 5 NTU and the time the water has been in contact with the chlorine is more than 30 minutes, the water is safe to drink from a biological point of view.

Chemical testing

The number of substances which can contaminate water is staggering, and in reality it´s impossible to test for all of them. Just stop and think how many different kinds of pesticides there are. Similarly, some metals and organic or radioactive compounds require highly specialized and expensive tests, which are often not available. For practical purposes, try and get the most complete test you can have done, and wherever possible, have it done at a public entity which specialises in water or health.

Excessive values

The values which exceed the limits can be corrected by treating the water, or by mixing water from various sources. If you have multiple sources to choose from, take the one which needs no treatment, even if it is further away and requires a greater initial investment. Water treatment is complicated and increases running costs.

3. 5 ENCRUSTIVE OR CORROSIVE?

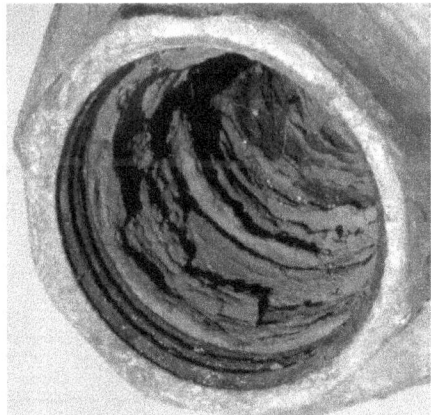

The build up of deposits due to encrustive water and tuberculation of corrosive water (see photo) can massively change the effective diameter and roughness of the pipe. It´s very important to be able to predict which of the two problems you are most likely to have to deal with.

To work out if the water has encrustive or corrosive characteristics, use the **Langelier Index**:

$$IL = pHa - pHs = pHa - ((9.3 + A + B) - (C + D))$$

Where:

pHa, pH of the water

A = (Log$_{10}$ [Total Dissolved Solids in mg/l] - 1) / 10

B = -13.12 x Log$_{10}$ (Water temperature in $^{\circ}$C + 273) + 34.55

C = Log$_{10}$ [Ca^{2+} in mg/l of CaCO$_3$] – 0.4

D = Log$_{10}$ [Alkalinity in mg/l CaCO$_3$]

If IL = 0, the water is in chemical equilibrium.

If IL < 0, the water has a corrosive tendency.

If IL > 0, the water has an encrusting tendency.

For practical purposes:

If the values are between -0.3 and 0.3, the water won´t give you problems.

Between -0.5 and -0.3, there´ll be some corrosion, but nothing too serious.

If IL< -0.5, corrosion will be a problem.

If IL > 0.5, there will be major deposits.

Problems with corrosion can be avoided by installing plastic pipe. Problems with encrustation can be avoided by reducing the pH of the water, or by selecting an alternative source, if possible.

Calculation example:

The water test from source A shows the following results: pH = 6.7; TDS = 46 mg/l; Alkalinity = 192 mg/l; Hardness CaCO$_3$ = 102 mg/l. If the water is at 12°C what precautions should I take?

The Langelier Index is used to work out whether the water is encrustive or corrosive:

A = (Log$_{10}$ [SDT] – 1)/ 10 = (Log$_{10}$ [46] – 1)/ 10= 0.066

B = -13.12Log$_{10}$ (T$^{\circ}$ + 273) + 34.55 = -13.12Log$_{10}$ (285) +34.55= 2.34

C = Log$_{10}$ [Ca^{2+} in mg/l of CaCO$_3$] – 0.4= 1.6

D = Log$_{10}$ [Alkalinity in mg/l CaCO$_3$]= 2.28

IL = pHa-((9.3+A+B)-(C+D) =6.7-((9.3+0.066+2.34)-(1.6+2.28))= -1.1

The water is highly corrosive. Pipes resistant to corrosion should be used, PVC or high density polyethylene (HDPE).

---

3. 6   PROTECTION OF SOURCES

To guarantee the safety of drinking water, the sources must be protected. On the one hand, animals and people must not be able to get close to the source. On the other hand, you need to prevent certain activities near the source which can pollute it. For most soils, a minimum of 30 meters is recommended between a potential source of contamination and the water itself. Sandy and rocky soils require more.

In the photo, the protection to stop cattle getting in and the hedge row in the background are in bad condition. The animals can get in, and although they can´t get to the actual spring, their faeces are dropped too close. To stop people getting in, a metal cage has been constructed around the spring.

Fig. 3.6. Protection of an artisan spring, Xhindaree, Somalia.

4. Pipe layout and topography

4. 1 GENERAL CONSIDERATIONS

In the layout the route taken by the piping is decided. To do this you´ll need the help of a general map.

Rights of way

Before deciding on a route, make sure you´ll be able to get permission to install the piping. You´ll often end up following the boundaries of private properties, even if this means using more piping. Another important consideration is making sure there will be access in the future for repairs and enlargements. Some areas, like cemeteries, can be highly sensitive and are not always that obvious.

Fig 4.1. Muslim grave.

Mines and military remains

Investigate very carefully the areas where the piping will go, to avoid accidents among the workers. Any area with remains of military hardware, like in the photo, are a bad candidate, due to the risk of explosive artefacts or unexploded ammunition.

Roads and paths

To avoid conflict, make it easy to find buried piping, stay clear of house constructions which may make repairs difficult, and to facilitate the transportation of materials, try to follow road and paths.

This can be tricky. The installation can damage electrical cables (see photo), conduits, telephone lines etc. Also, channels with waste water and latrine discharges often run along roads. Consult chapter 6 to see what to do when installing piping in special situations.

4. 2 TECHNICAL CONSIDERATIONS

High and low points

Sediment tends to collect in the low points and ends up strangling the pipe. This can be solved by installing washouts to empty the pipe, or by simply avoiding low points altogether.

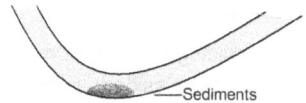
Sediments

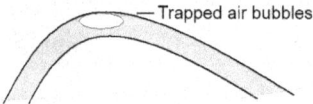
Trapped air bubbles

At the high points, air accumulates. This is more difficult to resolve, as air valves are expensive, logistically complicated and can cause a serious water hammer effect.

Distribution

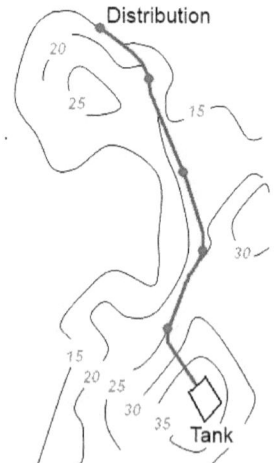

The best way of solving these problems is to design the pipe layout in such a way as to avoid points of inflection. Look at the layout in the diagram: from 35 meters at the tank, the pipeline drops rapidly to 25 meters. The idea is to pressurise the pipe as soon as possible. Once the 25 meter contour is reached, the pipe follows it, until it diverges from the main north-south direction and descends smoothly to the 20m contour. After this curve it descends to 15 meters. The result is that there are no points of inflection. Sediment is removed from the pipe at the end of the line and any air travels up the pipe and leaves at the tank.

Excavation work

The trench dug for the pipe can destabilise and weaken nearby structures: walls of houses, fencing, and electricity posts. The potential conflict caused by a house partially collapsing or an electricity post falling down is considerable.

Crossing rivers and streams

If you´ve designed a route following paths and roads, it´s simplest to cross rivers and streams by fixing the pipe to the bridge structure. If there are no bridges, you can submerge the pipe by weighing it down or make a cable bridge with steel cabling. This will all be dealt with later. In the case of streams, find the narrowest section for a suspended crossing, or the widest point if you´re going to bury the pipe.

Making bends

The sharper the bend the more energy is lost. Also, to compensate for the forces the pipe will be put under, concrete blocks must be put in place. The energy loss caused by accessories and how to retain forces with concrete blocks is dealt with further ahead.

To avoid having to build these blocks and waste time and energy unnecessarily with extra parts (elbows), you can use the margin of deflection allowed by the piping to smooth the curves. This is usually between 3° and 5°. For a 3° deflection and 6m pipes, the curve radius is 120m.

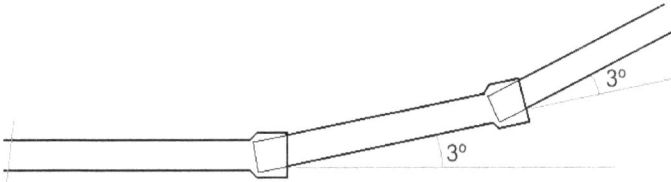

This margin means you can have bends with unusual angles. There are usually only 90°, 45°, and 22.5° elbows commercially available. All angles in between can be obtained by bending the pipe.

If the terrain won´t let you make a smooth curve and you need 34.6° all of a sudden, you can place two 90° elbows one after the other and twist them in the vertical axis.
Small diameter pipes and HDPE less than 110mm can be bent more sharply, to the point where they sometimes come in rolls.

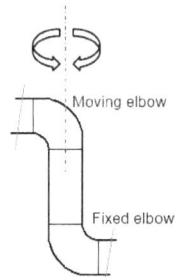

Moving elbow

Fixed elbow

Costs and complications

It´s very easy to plan a layout on paper which in practice is complicated and uneconomical. Avoid rocky areas, or areas with gravel or thick vegetation. Try and look for settled ground where you can avoid having to put in a sand layer and where you can use find locally available material.

Access

Installing piping correctly can require huge amounts of material, not just pipe, but also tons of sand to prepare the trench. Open terrain or areas of easy access are preferable.

4. 3   SITE MAPS

Getting hold of a map of the area is fundamental to begin planning alternative pipe routes.

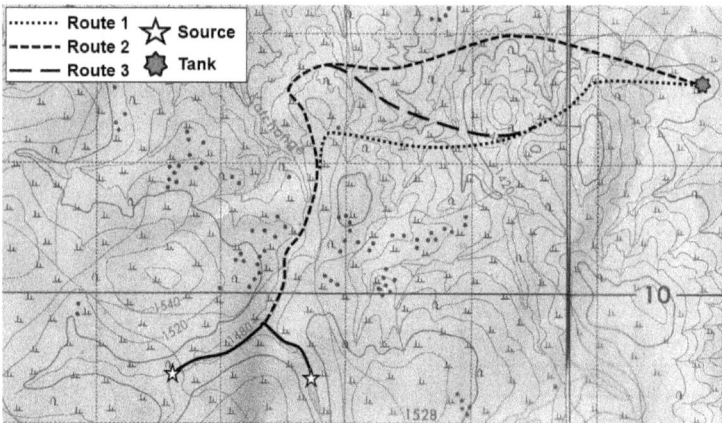

However, a topographic survey can almost never be replaced for its accuracy in measuring elevation. Note that the contour lines are in 20m intervals. 20m of pressure is often the difference between the maximum and minimum pressure of a system. On the other hand, trying to do a topographic survey without having a clear idea of the terrain can mean a lot more work is required.

4. 4   GETTING HOLD OF MAPS

Getting hold of a map can be very frustrating and take up a disproportionate amount of time. Less though if you don´t know where to look. They´re often in the most unexpected places:

In the project file!

It might seem obvious, but it´s often overlooked. Something in our subconscious makes us assume that there can´t be anything but obsolete information in that forgotten little room inside that rusted old locker. Maps, especially in the ex-colonies, are often relics from a long gone age, which are either no longer reproduced or are very difficult to find. That box full of ancient clutter is probably your best bet.

In old reports

For some strange reason, the survival rate for reports from the Pleistocene period is higher than that of essential information you need for any project. Look in the appendices and introductions to these reports.

On the walls

People love maps and plans. It makes them look serious. The map you´re looking for may well be on the wall of your boss´s office, or some other organization. Sometimes, truth is stranger than fiction. Following the Tsunami in 2004, the cartography agency of the UN told us they didn´t have any maps for the city of Meulaboh, while the exact map we were looking for was hanging on their wall with an aerial photo of the city to a scale of 1:10,000, and contours at 5m intervals. Most organisations had been working there for over 2 months without a single map of the city.

From official organisations

This is probably the first option you thought of. However, it´s often the one which takes most time and can give few or no results.

From businesses working in the area

Local businesses tend to have aerial photographs and endless maps of the area in which they work. If you want your pipe to follow the road, doesn´t it make sense to ask the people who are doing road maintenance work?

From the internet

If you have an internet connection, you´re in luck. You can access Google cartography for free. To do this, download the Google Earth program:
http://earth.google.co.uk/intl/en_uk/download-earth.html

Once you´ve installed and opened the program, you can navigate to the area of your project using the controls on the upper right hand corner:

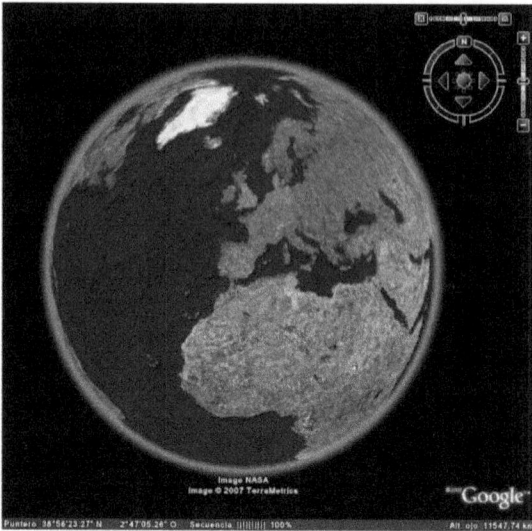

The vertical bar is to increase the zoom. Google will tell you when you reach images of maximum resolution. The horizontal bar lets you incline the view until it becomes 3D, which is really useful to visualize possible routes for the pipe:

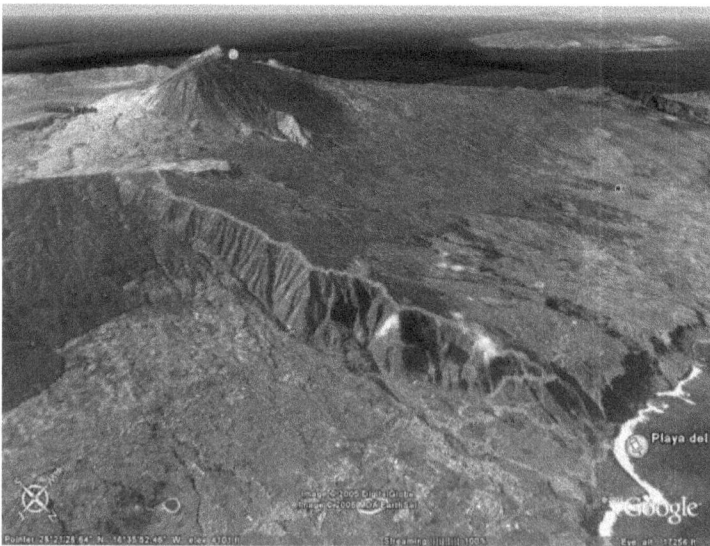

To be able to work quietly and offline, you'll need another program to download the images or use Google Earth Pro which is provided free by Google to NGO's.

GPS Software

Some programs for GPS, like CompeGPS, have the option to download existing maps of an area. They do this by connecting to NASA public servers, Agriculture Ministries, Google and other institutions, allowing you to choose which map you download.

Some of these programs can draw 3D images or obtain topographic profiles of the routes drawn out on the map:

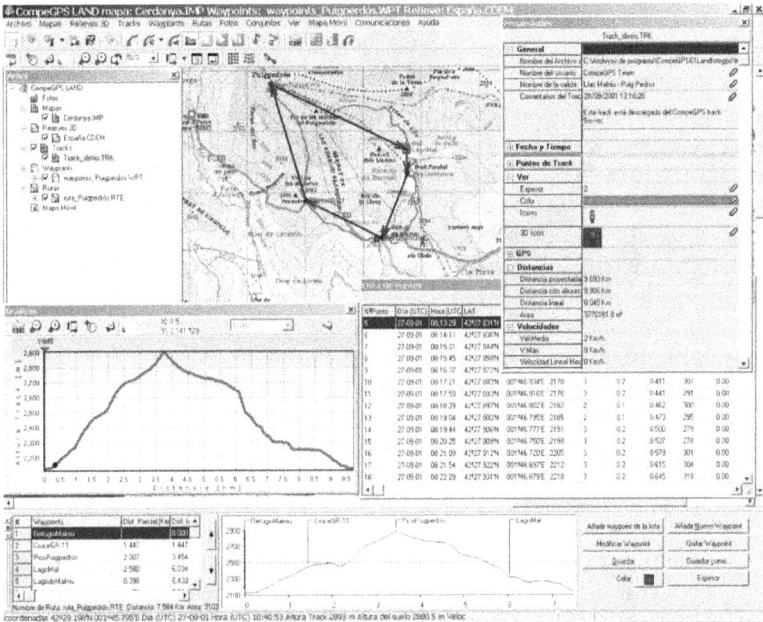

Fig. 4.4 Screenshot of CompeGPS with analysis of the topographic profile.

4. 5   GPS

A GPS is an instrument which tells you the coordinates of a given point with extreme precision (around 5m meters for conventional units). This is done by triangulation between a cloud of satellites.

Their main use is in taking marker or waypoints. For example, below you can see the marker points for the completely unknown system of Meulaboh. Superimposing the markers on the satellite image which was hanging on the wall, I could rebuild the map of the system:

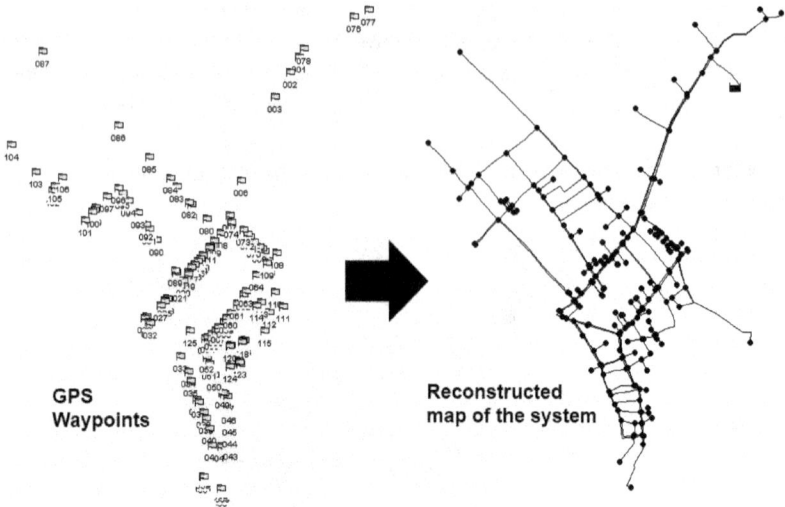

GPS
Waypoints

Reconstructed
map of the system

Common mistakes in using a GPS

With unit costs coming down, together with their ease of use and general popularity, they are ironically one of the most badly used and abused bits of kit you´ll find around. They can easily be used to wreck a project:

1. **Taking the altitude seriously**. Your GPS is going to show you a reading for altitude in meters (e.g. 1826m). If your GPS doesn´t have a barometer, the errors can be of several hundred meters, depending on the geometry of the satellites. If your unit has a barometer, keep your hat on, as the conventional models have an error margin of ±10m. To base your design on data taken with this margin of error is dangerous. With a range like that, it may mean that people who are at -10m receive their water at an uncomfortably high pressure, while those at +10m get nothing. Compare this to a topographic survey over several kilometres with an optical theodolite built way before you were born, and the error will only be a few centimetres. What´s more, for a survey you need 3 altimeters (section 4.7), not just one.

2. **Not selecting a Datum**. A datum is a reference point which has been used to construct the cartography of an area. The default datum of the GPS is WGS84, very useful in case you don´t have any maps, or for sharing information, but pretty inaccurate across the planet. That´s why maps are made with more local datum, which are shown very precisely on the scale, or underneath (Snowy Peaks, Arc 1960, etc.). Using a GPS with a datum

configured to WGS84, with a map that has a different datum, can lead to errors of several hundreds of meters.

3. **Not using UTM coordinates**. The longitude and latitude coordinates were invented for navigation with no obstacles, useful for sea and air travel. On land, the distance between 2 points or the position on a map can´t be determined without laborious calculations in which errors often creep in. Observe, for example, the important difference the comma and spaces have in each of the 3 expressions of measurement below:

-78.1947°	Degrees and decimals of a degree.
-78° 19.47'	Degrees, minutes, and decimals of a minute.
-78° 19' 47.00"	Degrees, minutes, seconds and decimals.

These 3 points are very far away from each other, as much as 14.5 km, as you can see in this image of Central America. If you´re not careful with the comma and spaces, you may find that when you get back from taking your marker points you´re in the Indian Ocean or a neighbouring country, or worst, your mistakes won't be so visible and you end up using them.

4. 6 TOPOGRAPHIC SURVEYS

In its simplest form, a topographic survey measures the difference in altitude and distance between consecutive points on a route. To do this, an indicator levelled exactly horizontal can be used together with 2 rulers to read off the difference in height.

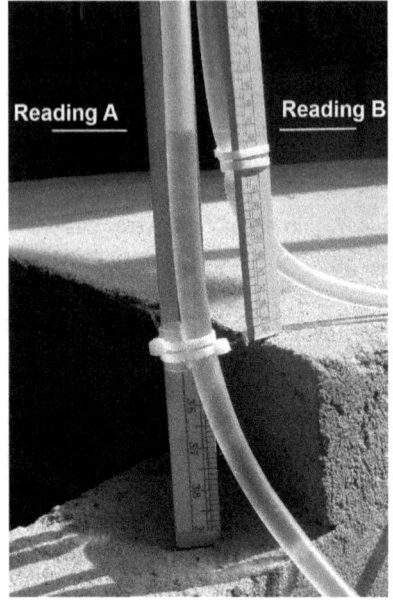

Reading A   Reading B

In the photo a water level can be seen. The water at each end of the pipe will settle at the same level, and the difference in height between each step is the measurement *A* less the reading *B* on each ruler. This kind of level is too slow though and the distances too short. Nonetheless, where no other equipment is available, it´s been used extensively to work out contour lines for agricultural terracing.

If the rulers are separated, a levelled sight (theodolite) can be used to determine the horizontal plane, and a tape measure used to measure the distance. You can work more quickly like this, and it´s the most common method.

For water supply we´re not really interested in the height above sea level, but more in the difference in height between points. This is why an arbitrary point is usually chosen (the source, the site of the future storage tank, the paved area in front of a school etc.) and all other points are measured with respect to this. This is the datum in the project maps.

It´s often possible to hire trained theodolite teams. This will result in profile diagrams which relate the cumulative distance to the elevation:

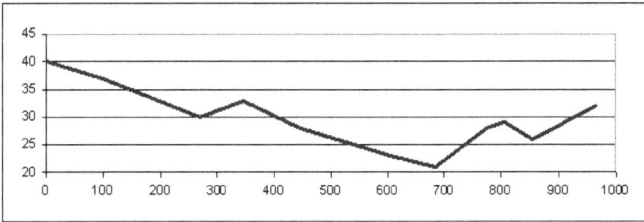

Altitude	40	37	32	30	33	28	23	21	28	29	26	32
Distance	0	100	120	50	75	100	160	80	90	30	50	110
Chainage	0	100	220	270	345	445	605	685	775	805	855	965

A   B   C   D   E

In this profile, the point B has an elevation of 28m, and a cumulative distance of 445m, being 100m beyond the previous point. Swinging the theodolite round to each respective ruler gives the 2 angle measurements. Its subtraction shows the angle with the change of direction and can be shown as a floor plan:

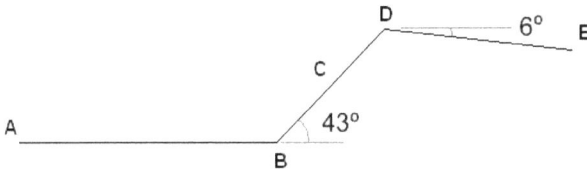

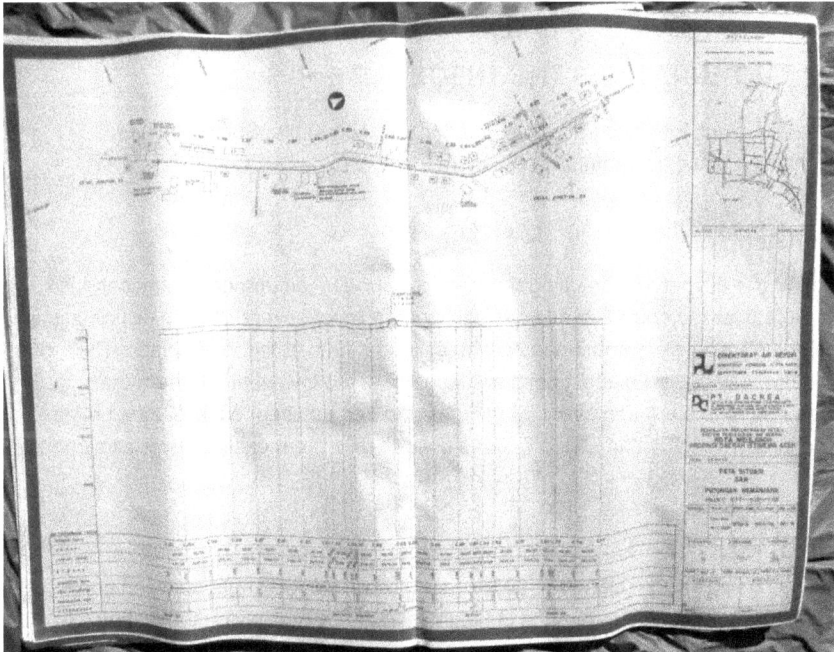

Fig. 4.6. Rapid digitalisation of a topographic survey by taking a photo.

Practical questions

1. Decide and mark the project datum. This datum will be used for all future measurements and enlargements, which means it needs to be lasting and easily recognisable. Signal it with a marker, with the elevation inscribed.

2. On the day you did a superb survey and the pipes only just arrived 5 months later. Ummmmm....where was it we measured? Mark out the route with wooden sticks and paint or record them in your GPS. Place a stick every 500m with the elevation and distance.

3. To check the precision, measure in both directions, up and down, and compare the measurements. They should be within a meter of each other. You can also cross check measurements by comparing different routes. If the start and end points are the same, the difference in height- whatever your route- must be the same.

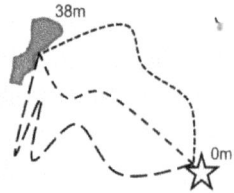

4. Don´t survey over terrain you can´t install the piping: bedrock, flooded areas, gravel...

4. 7 ALTERNATIVES THE THEODOLITE

Generally much less accurate, but nonetheless acceptable. Some of these alternatives could be useful at some point.

Altimeter

With quality altimeters you can get decent results. The difference in atmospheric pressure over the course of one day can give errors of several meters, which is why you need 3 altimeters and 3 people who calibrate them in the same place. Then, one altimeter stays at the highest point and another at the lowest along the route. Measurements are taken every 30 minutes and can be used to correct the pressure variations. The third altimeter is the one used to take the measurements all along the route.

Abney Level

This is a device with a sight that allows you to work out the angle between horizontal and a target point. The difference in height is found by multiplying the distance by the sine of the angle measured.

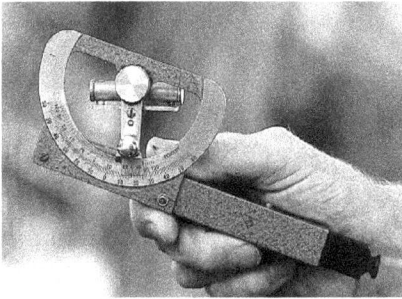

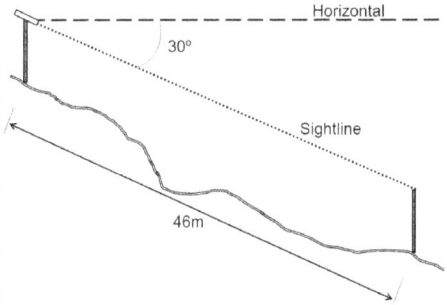

If the angle measured is 30° and the distance is 46m, the difference in height is:

46m * sine 30° = 46m * ½ = 23m.

Most measuring instruments are designed with a precision down to half the minimum unit. Even if you can read off with more precision than that, the device won´t be able to keep up with you.

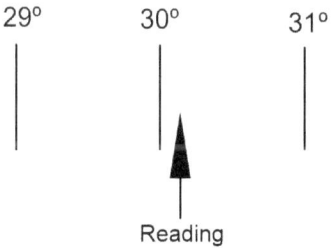

29°   30°   31°

This reading is 30° although if you squint you could say it´s 30.15°. The precision is half a degree and you´re choosing between 30° and 30.15 °. As 30° is closer that´s the correct reading.

Reading

5. Basic designs

Most common designs and many of the weirdest ones are made up of combinations of some of these basic scenarios:

5. 1 FISHBONES

This is the most common, in which the pipe work is branched:

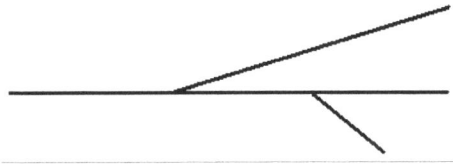

To do the calculations you can start at the source downwards, which is usually easier. At each mainline you add the flows of the minor pipe which branches off:

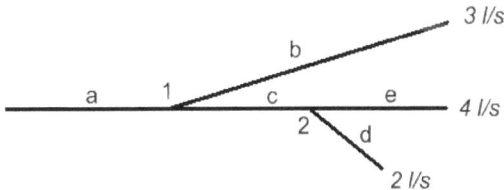

Nodes 1 and 2 include 5 pipe reaches. The pipes *b, e* and *d* only have to deal with their final demands, 3, 4 and 2 l/s respectively. So that the water reaches the pipes *e* and *d*, pipe *c* has to carry both their flows, 6 l/s. Pipe *a* needs to carry a flow of 9 l/s.

Calculation example:

Calculate the system proposed in the text for HDPE with a pressure at the taps of 1.5 and 3 bar, with this topographic data:

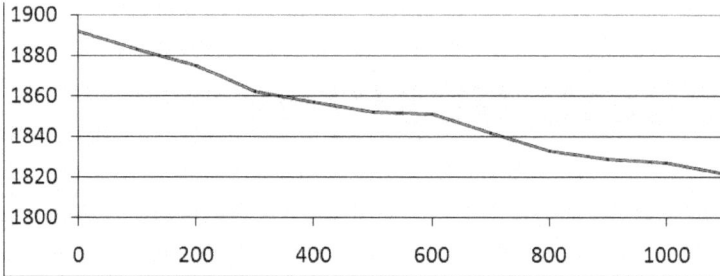

H.	1892	1883	1875	1862	1857	1852	1851	1842	1833	1829	1827	1822
Long	0	100	100	100	100	100	100	100	100	100	100	100
L. Acum.	0	100	200	300	400	500	600	700	800	900	1000	1100

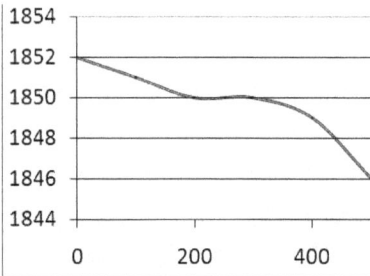

1   2   F2 (4 l/s)

H.	1852	1851	1850	1850	1849	1846
Long	0	100	100	100	100	100
L. Acum.	0	100	200	300	400	500

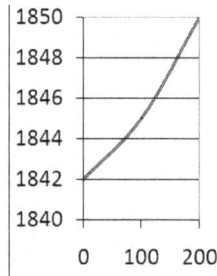

1   F1 (3 l/s)

H.	1842	1845	1850
Long	0	100	100
L. Acum.	0	100	200

2   F3 (2 l/s)

REACH A

Pipe *a* has to carry 9 l/s. At point 1, the elevation is 1852m, very close to that of tap F1 (1846m) and F3 (1850m). We need sufficient residual pressure for the branch lines. 10m won´t be enough. Tentatively, we can aim for 25m of pressure, which will allow a head loss of:

$$1852m + 25m \text{ of pressure} = 1877m$$

$$(1892m-1877m)/0{,}500km = 30m/km$$

In the tables, there isn't a value close enough to 9 l/s for <u>90mm</u> pipe:

20.00	6.236	1.27
30.00	7.798	1.58
45.00	9.740	1.98

Although it doesn't follow an exactly linear relationship, the intervals between the data are sufficiently small for you to do a linear interpolation. The generic formula is:

$$\frac{J_x - J_{inf}}{J_{sup} - J_{inf}} = \frac{Q_x - Q_{inf}}{Q_{sup} - Q_{inf}}$$

Where: J_x, the head loss you are looking for.
J_{inf}, J of the lower flow rate.
J_{sup}, J of the higher flow rate.
Q_x, the flow rate in question.
Q_{inf}, the lower flow rate.
Q_{sup}, the higher flow rate.

$$\frac{J_x - 30}{45 - 30} = \frac{9 - 7.798}{9.74 - 7.798} \rightarrow J_x = 39.28 \text{m/km}$$

The dissipated energy will be: 0.5km * 39.28m/km = 19.64m

The residual pressure at 1 is: 1892m - 1852m – 19.64m = 20.36m

20.36 meters is close enough to the 25 meters mentioned originally.

REACH C
Pipe c has to carry 6 l/s in total over a distance of 200m. Keeping the pressure at 20m at point 2, the water will have sufficient pressure to climb up the branch line d.

J_{max}= (20.36m + 1852m – 1842m - 20m)/0.2 km = 51.8m/km

There's no pipe that's near this value. However as the distance is small and we don't need exactly 20m (22m or 24m are equally valid), we can use <u>90mm</u> pipe. For 6.2 l/s J= 20 m/km. The pressure are point 2 will be:

P_2= 20.36 +1852m - 1842m - (20m/km*0.2km) = 26.36m

REACH E

Pipe *e* has to carry 4 l/s in total over a distance of 400m with a residual pressure of between 1.5 and 3 bar (15-30m). You need to find a pipe with a head loss between those values, for 4 l/s:

$$J_{min}= (26.36m + 1842m - 1822m - 30m)/0.4 \text{ km} = 40.9 m/km$$
$$J_{max}= (26.36m + 1842m - 1822m - 15m)/0.4 \text{ km} = 78.4 m/km$$

Looking at the tables for 63mm pipe:

45.00	3.752	1.56
60.00	4.396	1.82

For 4 l/s the value will be between 45m and 60m, or even 40.9m and 78.4m.

To work out the residual pressure we look for the right value by interpolating:

$$\frac{J_x - 45}{60 - 45} = \frac{4 - 3.752}{4.396 - 3.752} \qquad J_x = 50.78 m/km$$

This value is between the previous ones. The residual pressure will be:

$$P_{F2}= 26.36 + 1842m - 1822m - (50.78m/km * 0.4km) = 26.04m$$

The mainline is therefore:

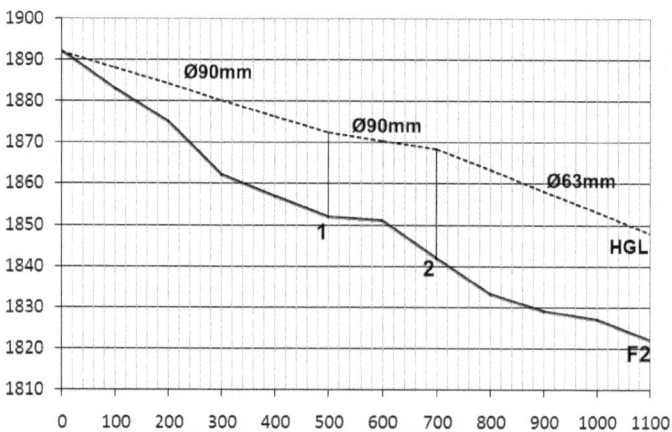

REACH B
Pipe *b* has to carry 3 l/s over a distance of 500m, to an elevation of 1846m, leaving from 1852m. The pressure at 1 has been worked out as 20.36m.

$$J_{max}= (20.36m + 1852m - 1846m - 15m)/0.5 \text{ km} = 22.72m/km \text{ or less.}$$

Looking at the tables for a pipe of 90mm, we find a value of 5.5m/km. Check that this doesn't exceed the maximum pipe pressure:

$$P_{F1}= 20.36 + 1852m - 1846m - (5.5m/km*0.5km) = 23.61m$$

In case this has been exceeded, a combination of pipe diameters would have been installed, as seen in section 1.6.

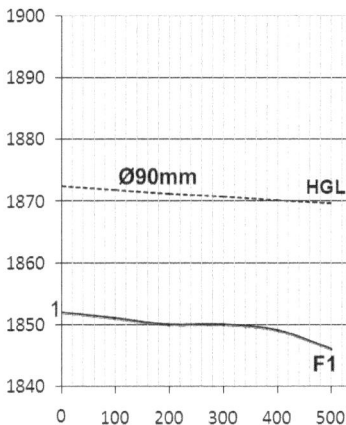

REACH D
Pipe *d* has to carry 2 l/s over a distance of 200m, to an elevation of 1850m, leaving from 1842m. The pressure at 2 has been worked out as 26.36m.

$$J_{max}= (26.36m + 1842m - 1850m -15m)/0.2 \text{ km} = 16.8m/km \text{ or less.}$$

Looking at the tables for 63mm pipe, we find the value of 15m/km. Check the maximum pressure is not exceeded:

$$P_{F3}= 26.36 +1842m - 1850m - (15m/km*0.2km) = 15.36m$$

The partial diagram looks like this:

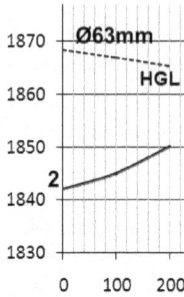

With all the branch lines on the same diagram:

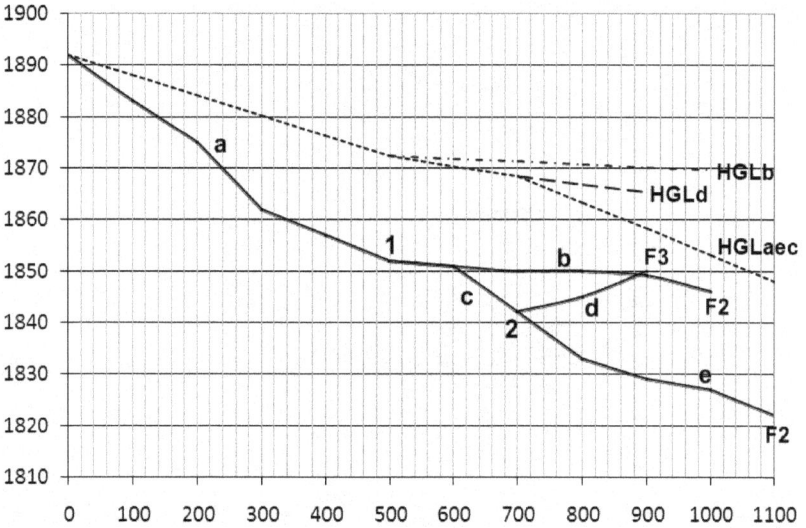

IMPORTANT: Note that when there's no demand, at night for example, the pressure at point F2 is 1892m-1822m=70m. This kind of pressure at a tap is dangerous and useless to people. Read carefully section 5.3 to find out how to solve this.

5. 2 MULTIPLE SOURCES

There are times when it´s necessary to use multiple sources, which are unlikely to be at the same height. When sources are joined at differing pressures, the higher pressure interferes with the flow out of the lower.

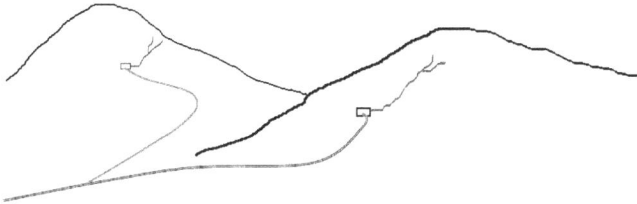

In these situations, **both sources need to meet each other with the same residual pressure.** When there is no demand, the higher source can discharge into the lower one. To avoid this, a non-return or check valve is installed in the pipe of the lower source.

Calculation example:

On one side of a valley a source has been tapped giving 4 l/s at an elevation of 60m (north source). At 52m on the opposite side, a second source provides 2 l/s (south source). Both flows are to be joined at 37m of elevation with HDPE pipe. With the following topographic survey at hand, which pipes should be installed?

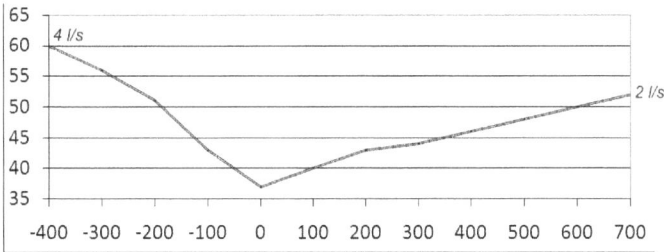

Elevation	60	56	51	43	37	40	43	44	46	48	50	52
Distance	0	100	100	100	100	100	100	100	100	100	100	100
Chainage	-400	-300	-200	-100	0	100	200	300	400	500	600	700

North S.   Junction   South S.

We want the pressure at the junction to be greater than 10m, although the topography won´t let us pressurise the system much more than that.

The pipe with less room to manoeuvre comes from the south source. Let´s start with this. The maximum head loss is:

$$J_{max} = (52m - 10m - 37m)/0,7km = 7.14m/km$$

For a flow of 2 l/s, the pipe which leaves us enough pressure is 90mm. The pressure at the junction is:

$$P= 52m – 37m - (2.75m/km*0.7km) = 13.07m$$

The required head loss for the North pipe is:

$$J_{north\text{-}union} = (60m - 13m - 37m)/0.4km = 25m/km$$

This section consists of a mixture of 63mm and 90mm pipe. To find the head loss for 4 l/s in the 63mm pipe, we need to interpolate between the values of 45 and 60, as in the previous exercise.

$$\frac{J_x - 45}{60 - 45} = \frac{4 - 3.752}{4.396 - 3.752} \rightarrow J_x = 50.78m/km$$

To find the length required for each pipe, the same procedure is followed as in exercise 1.6:

Head loss for X km of 63mm pipe – Head loss for remaining distance of 90mm pipe = Maximum possible drop

X * 50.78 m/km + ((0.4km –X) * 9m/km) = 60m-37m-13m

50.78X + 3.6 - 9X = 10 → 41.78X = 6.4   → X = 0.153km

Note that the order in which the pipes are placed is important. If you put the smaller diameter pipe first the hydraulic grade line will drop below ground:

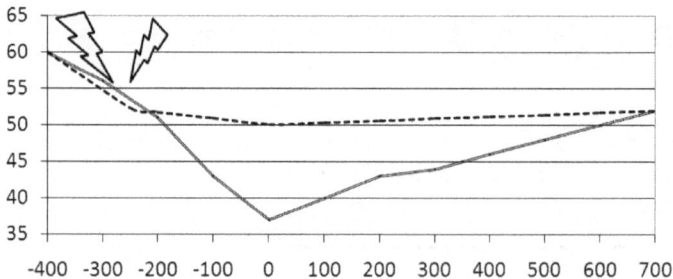

The grade line and topographic graph then look like this:

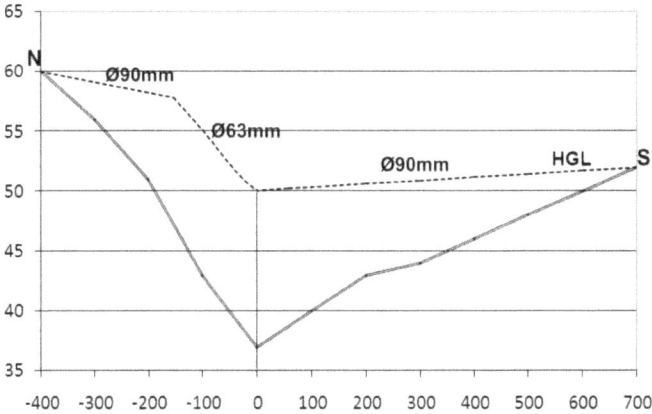

The hydraulic grade lines of each pipe reach the junction with the same pressure, 13m.

5. 3 EXCESSIVE DROP

The minimum pressure ensures that the users get water and don´t despair with a miserable drip. It´s just as important not to exceed the maximum limits. With excessive pressure the systems become dangerous, fragile, and full of leaks. A huge amount of water ends up being lost and wasted.

On the following page, you can see what happens when you try to fill a bucket with 3 bar of pressure. Note that two hands are needed to open the tap. The jet of water coming out of the tap gets the entire surrounding area wet, including the feet of the users, and once the tap is shut, the bubbles fade away to leave a half full bucket. The system becomes wasteful, dangerous, unpleasant and inaccessible for the greater part of the population, if automatic shut-off taps are used.

In areas of low lying houses, try and make sure your system doesn´t exceed 25m of pressure at any moment.

To reduce the pressure, **break pressure tanks (BPT)** can be installed, when the difference in height between the source and the user exceeds 25-30m. A BPT is a small tank which the water discharges into. As this happens, pressure is lost, returning to atmospheric pressure, the same as when a tyre is punctured.

Fig. 5.3.b Excess pressure at a public tap, Lugufu, Tanzania.

Fig. 5.3. Excess pressure caused by topography. BPT, Qoli Abchakan, Afghanistan.

A BPT allows the pressure to return to 0. By choosing an installation pressure you can control the maximum available pressure in the system. As in other points in the system, the entrance to the BPT should have a minimum pressure of 10m to avoid surprises, and there should be more water coming in than leaving the tank. In section 12.3 the workings and construction of BPT are explained.

Calculation example:

Work out a gravity flow distribution in PVC so that the only point of demand (1 l/s) registers pressure between 1 and 2.3 bar at all times:

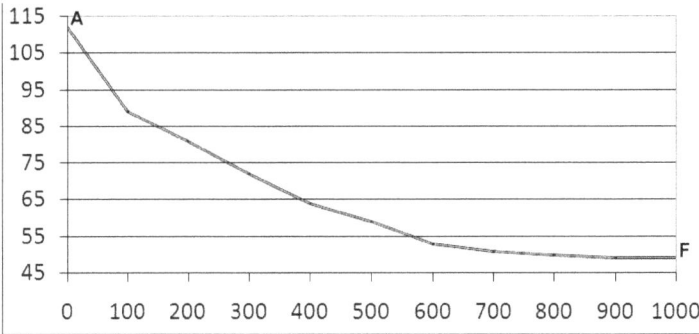

H.	112	89	81	72	64	59	53	51	50	49	49
Long	0	100	100	100	100	100	100	100	100	100	100
L. Acum.	0	100	200	300	400	500	600	700	800	900	1000
	A										F (1 l/s)

The pressure at times of no demand is: 112m - 49m = 63m

To respect the design maximum, the highest possible elevation for the BPT is:

C_{max} = 49m + 23m = 72m

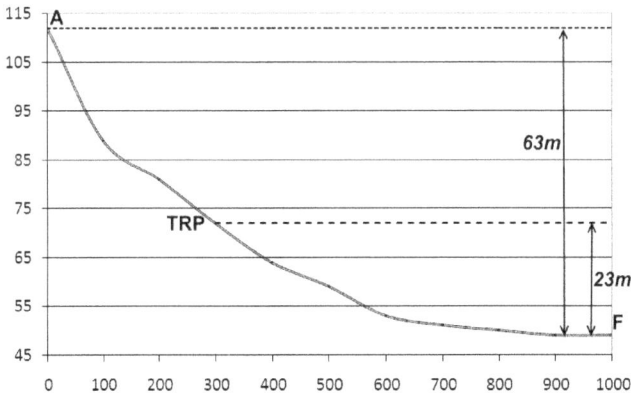

Placing the BPT at 72m, the pipeline is divided into one part *A-BPT* of 300m, and another *BPT-F* of 700m. The pipe which covers this last section can tolerate a head loss of:

$$J_{max} = (72m - 49m - 10m) / (0.700km) = 18.57m$$

Looking at the tables of PVC for a flow of 1 l/s, we find a value of J=3.75m/km for *63mm* pipe. The pressure at *F* will be:

$$P= 72m - 49m - (3.75 m/km * 0.7km) = 20.37m$$

The first section has to be worked out. To guarantee the supply, the pressure arriving at the BPT has to be at least 10m. However, with more pressure, the float valve will have problems closing when there is less demand downhill from the BPT. The same flow of 1 l/s is used.

$$J = (112m - 72m - 10m) / (0.300km) = 100m/km$$

Note that this value is outside of any of the tables. What's more, it would mean installing very small diameter pipes which are easily blocked. The solution is to install a pipe one size up, 40mm, and strangle the inlet to the BPT with a semi-closed valve. This valve will take care of all the head loss the pipe couldn't.

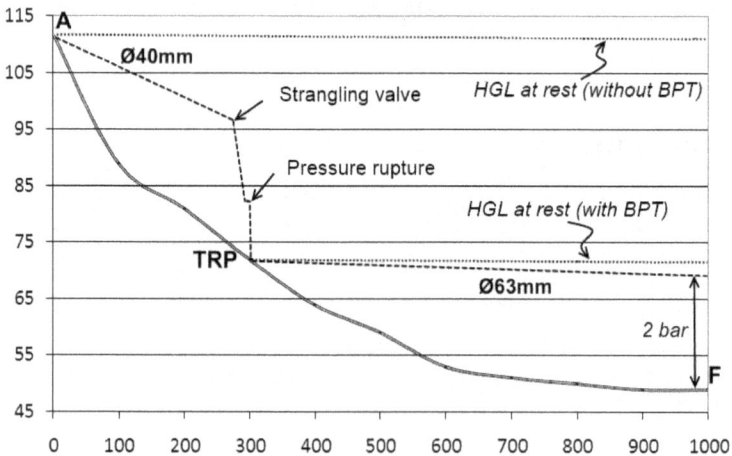

Note the big differences in pressure between the moments of no demand, with demand and with no BPT.

5. 4 VALLEYS

Up to now, we´ve ignored higher pressure pipes, PN16. They are very useful when the pressure in a pipe is too great. If the maximum design pressure at the taps is 30m, this doesn´t mean that at points along the way the pressure may be much more than that. Have a look at this real-life example:

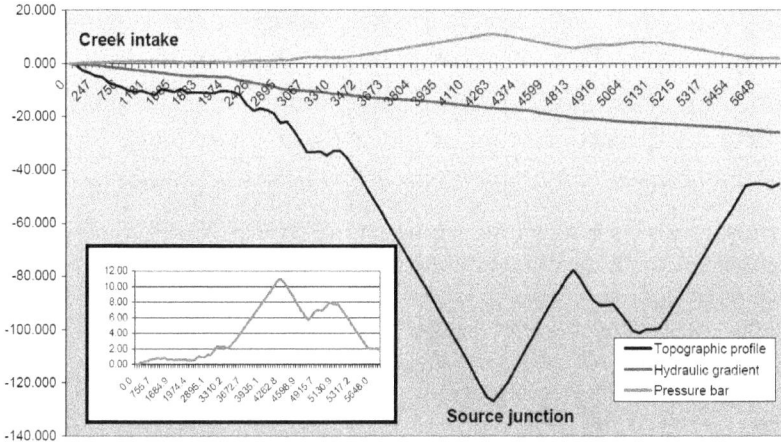

The pressure at the "Source Junction" is 12 bar, excessive for a pipe with a maximum limit of 10 bar. In fact, **never go over 80% of the specified capacity of a pipe** in your design.

When the maximum pressure in a PN10 pipe, being 10 bar* 0.8 = 8 bar, is reached you need to change to PN16. With PVC, the pressures are often expressed in classes:

Class B, 6 bar
Class C, 9 bar
Class D, 12 bar   12 bar * 0.8 = 9.6 bar maximum
Class E, 15 bar   15 bar * 0.8 = 12 bar maximum

For reasons of mechanical resistance, don´t install class B or C pipe.

With HDPE, the pressure rating is often displayed with a line of colour running along the pipe:

Red   6 bar
Blue   10 bar
Green   16 bar

5. 5 PRESSURE ZONES

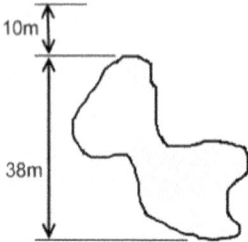

Imagine a system where the source is 10m above the highest user. Between this and the lowest point there's 38m. With the water at rest, the lowest user has 48m of pressure, almost 2 bar above the maximum. To resolve this problem, the area can be divided in two:

Zone *A*, covers the first 20m, fed directly from the source. Zone *B* covers the following 18m, passing first through a BPT placed 10m above the highest user in B. In each zone the minimum pressure will be 10m and the maximum 30 and 28m respectively.

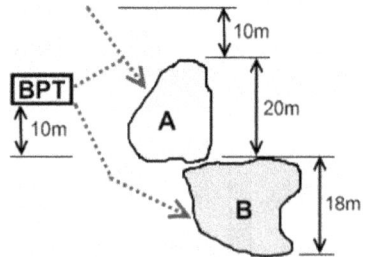

It's fairly common that in the zone you're dealing with, the difference in elevation between the high and low points is too great. To avoid making too many zones, a useful rule of thumb is this: make only enough zones to remain within a 35m pressure range.

Separating the pressure zones doesn't require that much more pipe. In areas where there's not much room to install a BPT, a pressure reducing valve can be used instead.

5. 6 TWISTED ROUTES

All accessories produce turbulence which increases friction. Normally these losses are very small and don't need to be taken into account for the calculations.

At other times, the route is so twisted or the height difference is so limited that they need to be taken into account to ensure proper functioning.

Minor losses

These are energy losses which are
produced by turbulence, introduced by
everything that´s not a straight pipe: elbows,
valves, reductions, T´s etc. In the photo, a
half closed valve creates conical turbulence
coming out of a tap. This can be seen
clearly in the shadow on the wall and
against the blue fabric.

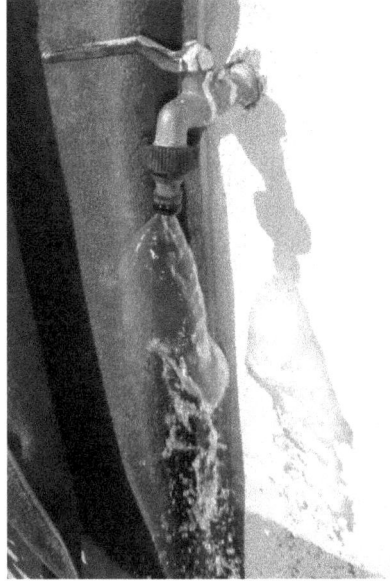

Minor losses are more critical in smaller
pipes (<25mm) and in those where the
water flows at high velocities.

Calculating minor losses

To calculate use this formula:

$$h = K\left(\frac{v^2}{2g}\right)$$

Where: h, head loss.
v, velocity in the pipe (see tables in Appendix B).
g, gravitational constant, 9.81m/s^2.
K, head loss coefficient depending on the accessory.

Some approximate values can be found in the table below:

FITTING	K COEF.
Ball valve, fully open	10
No return valve, fully open	2.5
Gate valve, fully open	0.3
90° elbow	0.8
45° elbow	0.4
Return elbow (180°)	2.2
Standard ' T ' (run-run)	0.6
Standard ' T ' (run-side)	1.8
Abrupt entrance	0.5
Abrupt exit	1

You can find more approximate values in section 2.5, reference 19, and in appendix
16, reference 5. For exact values, consult the manufacturer.

Calculation example:

Calculate the head loss of a 90° HDPE elbow, 90mm, with a peak flow of 4 l/s.

In the tables of Appendix B, 4 l/s corresponds to a velocity of 0.81 m/s. The coefficient of a 90° elbow is 0.8.

$$h = K\left(\frac{v^2}{2g}\right) = 0.8\left(\frac{0.81^2}{2*9.81}\right) = 0.027m$$

3 cm of head loss is negligible. At such low velocities, it´s not usually necessary to do the calculation.

5. 7 LOOPED SYSTEMS WITHOUT A COMPUTER

Bad news! Even the simplest system with only one loop can´t be designed the way we´ve been doing it up to now, as the water can reach one point via different routes.

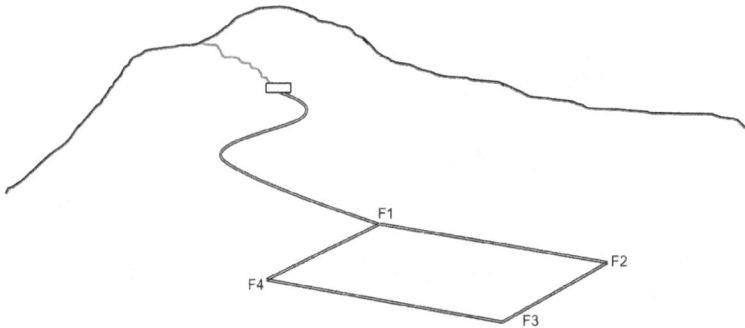

To design these systems by hand the Hardy Cross method is used. You can find the theory and a worked example in *How to Design a Gravity Flow Water System (Arnalich, 2009)* Even though it requires pre-University level maths, in reality it´s more laborious than complicated for simple systems.

Nowadays it´s a method that´s no longer in use, which is studied more as a theoretical exercise than for practical application purposes. Here I´m assuming you have access to a computer and basic computer skills.

Learning to use a design program to "Hardy Cross" level will take you much less time, 3 or 4 hours, and will be much more satisfying. This is the subject of the next chapter.

6. Water and the computer

6. 1 PROGRAMS

There are two free programs with distinct approaches. For its power of calculation, its gentle learning curve, its universality and absence of limitations, the following chapter will focus on the second, Epanet.

Neatworks

This is a program developed by the NGO *Agua para la Vida* (specialists in gravity flow water projects) and Logilab. It can be downloaded in Spanish, English and French here:

http://www.ordecsys.com/neatwork/

The program is limited to gravity flow projects. One of the interesting things is that it assists design and has a cost optimizer.

Epanet

This is a program from the North American Environmental Protection Agency, a worldwide reference in system calculation. It´s power of calculation is used as a major alternative to commercial programs. It includes a simple how-to guide and is quick to learn. It´s available in Spanish, English, French and Portuguese. The links to each version can be found here:

www.epanet.es/descargas.html

The instructions which are provided here are for the English version. Unless you want to use another version, download this one.

6. 2 GETTING TO KNOW EPANET

As you can see EPANET doesn't look too intimidating:

Fig. 6.2. Screen shot of Epanet. Calculation of the Meulaboh system, Indonesia.

The best way of exploring Epanet is to use the excellent introductory tutorial which is included in the Help. It'll take you one or two hours to finish it. Once you're done, you'll know enough to be able to follow the instructions below.

Introductory tutorial

Once you've downloaded, install the program. The installation is simple.

The tutorial can be found by clicking on the Help menu, and selecting "Tutorial" in the drop down menu:

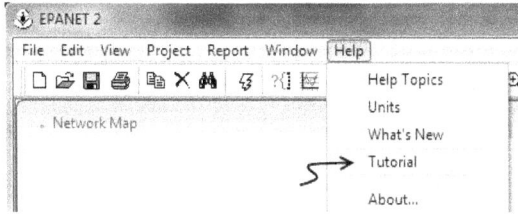

If you are using Windows Vista or Windows 7 operating system, you´ll get an error message. To be able to see the help files you can download and install this application (at your own risk):

www.epanet.es/descargas/ayudaVista.zip

You´ll be able to open a new window, which you can navigate by clicking on the arrows. Follow the instructions until you´ve completed the exercise:

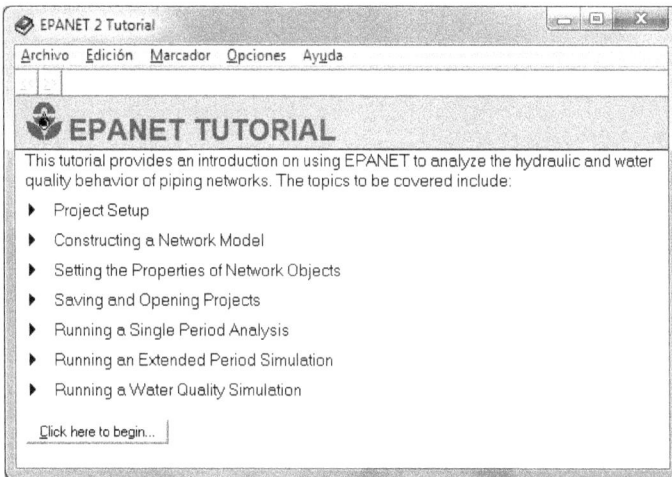

6. 3 CONFIGURING EPANET

For the calculations to make sense, you need to do some simple initial configurations:

1. Changing the units to LPS

In the tutorial you'll have seen that Epanet doesn't include the units in most of the windows or in the representation of the results. That's why it's fundamental that you get to know them. The most comfortable configuration is LPS (Litres Per Second).

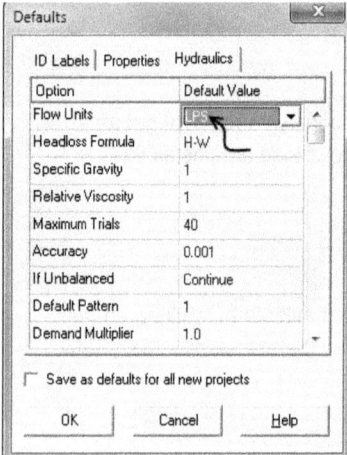

Click on *Project*, then *Defaults* in the drop down menu, and click on the *Hydraulics* tab. There, click on *Flow Units* and select *LPS* from the list.

With this change, the units are:

- Flow: litres/second
- Pressure: meters (column of water)
- Diameter: millimetres
- Pipe length: meters
- Elevation: meters
- Dimensions: meters

2. Selecting a calculation formula:

The most simple and direct is the Hazen-Williams, H-W in the window.

With this formula, the friction coefficient for the pipe or roughness is:

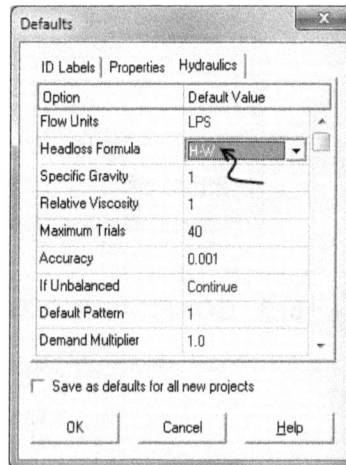

HDPE, PVC	140-150
Galvanized iron	120
Cast iron	130-140

To choose the calculation formula, in the same window and tab as the previous one, change the *Head loss Formula* to H-W.

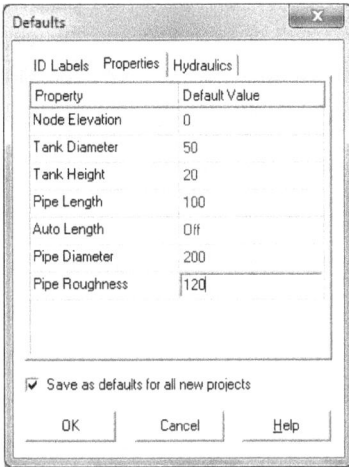

3. Selecting the default values

Selecting some default values can save you time, and means you don't have to go changing the properties of one pipe or node one at a time. If your system is regular and repetitive, you can select a default pipe length. If you think that most of your pipes will be 90mm, the same applies. A value which you must include is the pipe roughness.

To do this, click on the *Properties* tab and enter in the desired value. In the image, all pipes are 100m long, 200mm diameter and with a roughness of 120.

In Epanet, use internal pipe diameters. Check Appendix B for the approximate values for plastic pipes.

6. 4   GETTING OVER YOUR FEAR WITH AN EXCERCISE

The small town of Massawa has needed a water system for some time. Traditionally the water has been carried by donkey, from a stream 6km away, but miraculously a spring has been found on a nearby hillside at an elevation of 36m, and the town is excited. The estimated flow is 3 l/s. The system is planned to supply 6 public tap stands, all at an elevation of 17m, except number 6 (22m) and number 1 (25m), according to the map below.

Distances:
Spring-1   800m
1-2   400 m
2-3   300 m
3-4   250 m
3-6   500 m
5 to pipe 3-4   200 m

Design a PVC system capable of supplying 0.2 l/s to each tap and 1 l/s to the school.

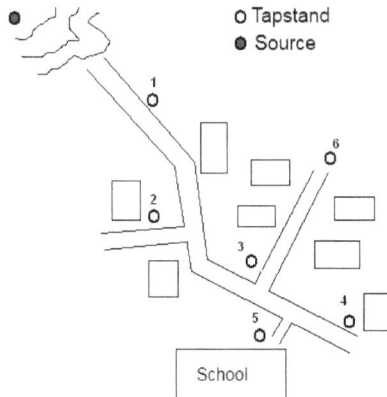

1. Click on *Project* and then *Defaults.* Under the *Hydraulics* tab in the window that opens, select the Hazen Williams (H-W) *Head loss Formula,* and change the units to *LPS.* Configure the *Defaults* in such a way that you save yourself work. Either way, make sure you enter the *Pipe Roughness* as 140 under the *Properties* tab (PVC).

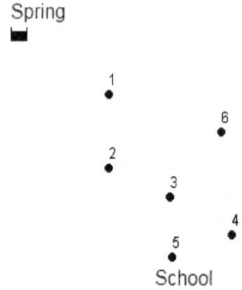

2. Draw the system nodes using the bar with these figures:  . Node 5 connects at some point to the pipe which runs between 3 and 4.

3. To do this, you'll need to draw in a node with no demand, which in reality is a "T". In the image below you can see this T before burial, and the corresponding extra node in the Epanet diagram.

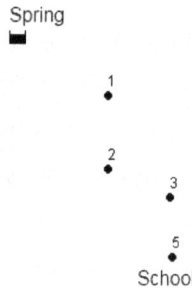

4. Join the nodes with pipes in the most logical way you see fit. In this case, it makes most sense to follow the route of the main road. The length of reach 3-4 has to be divided in two, roughly according to the proportions in the diagram or in real life. For example, we can make the left-hand pipe 150m long, and the right-hand pipe 100m, to give a total of 250m.

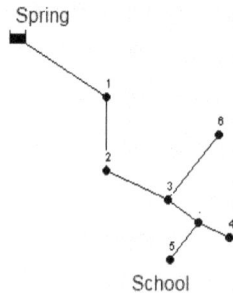

5. Introduce the data for elevation, length, demand and friction where the values are not predetermined. To change the properties of an object, double click on it.

6. Calculate the system by clicking on 🜀 . The most likely is that you´ll see a window come up which says *Run was successful*, which means the pipes are sufficiently sized. But watch out, *sufficiently sized* can be anything between the minimum diameter which will work and one the size of the Solar System!

7. Change the scale of the legend scale so you can comfortably see the results. To open the window that lets you do this, right click on the legend. If you double left click on it, it´ll disappear. To bring it up again, click on *View*, *Legends,* and then *Nodes:*

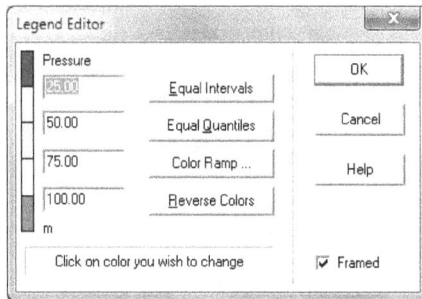

Epanet comes with a predetermined scale. Although it´s not all that helpful, you can change it. Imagine for a moment that in your system, you take the correct pressure ranges to be between 1 and 3 bar (10 and 30m). In this case, the best change you can probably make is to leave as dark blue the points with negative pressure, followed by a range of points with low pressure, from 0 to 10 meters, with the design margin in another colour, 10-30m.

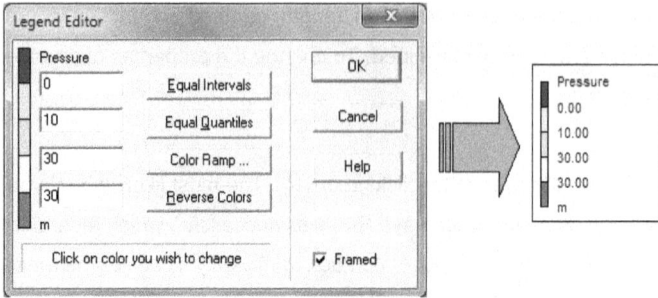

In this legend we´ve ignored the additional yellow scale. Once you´ve changed the diagram legend the screen will be updated like this:

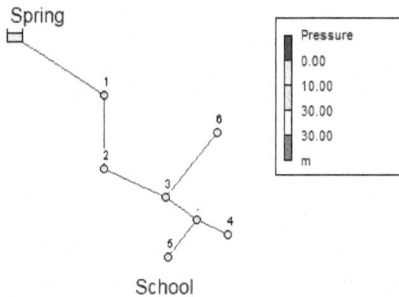

This system delivers all the right pressure values. But be careful! The job´s not done yet. Try changing the pipe from the spring to tap 1 with a pipe which has a diameter of 1km (1,000,000 mm). The system still works, even though it´s completely absurd.

The *Run was successful* window is only an invitation to continue optimizing the system, and is by no means Epanet telling you your system is ready.

8.   This means you need to reduce your pipe diameters down to the minimum, such that the pressure at all points in the system are over 10m. The first changes, as an example, are detailed in the points below. But before that, take note of this:

The most logical is to start from the spring and work your way down. If you start with the points furthest from the source, you´ll soon realize that any changes in pipe diameter you make later, nearer to the spring, will change everything you´ve meticulously worked on further downstream. You´ll end up in a spiral of never ending modifications.

9. Change the diameter of the pipe from the spring to tap 1, to 75mm and click on *Run*. Tap 1, with 8.26m of pressure, doesn´t meet the design minimum of 10m of pressure.

10. Change the diameter to 100mm. With 10.33 meters of pressure, you can take it as the correct pipe diameter.

Note that we´ve not used diameters such as 92.319mm, which would leave the pressure at point 1 at exactly 10m. Don´t waste time making adjustments like this and only use the pipe diameters which are commercially available in your area. Remember that when you order pipe, the diameter you´ve worked with in Epanet is the internal diameter. That means if you´re working with an internal diameter of 75mm for an HDPE pipe, you need to order 90mm pipe, as the 15mm of difference corresponds to the thickness of the pipe walls.

11. If the pipe from Tap 1 can be 100mm, it´s likely that the rest will be 100mm or less. If this is not the case, we´d be creating a bottle neck below the tap, and this is only done in special circumstances.

12. Carry on reducing the pipe diameters until you´ve optimised the system. There´s never just one right answer, just various possible solutions.

Remember to click on *Run* after making a series of changes so that Epanet takes them into account. If you don´t, you and Epanet may be working on very different systems!

This could be one possible solution:

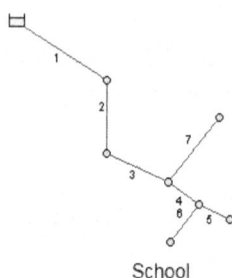

Link ID	Length m	Diameter mm
Pipe 1	800	100
Pipe 2	400	75
Pipe 3	300	75
Pipe 4	150	50
Pipe 5	100	18
Pipe 6	200	50
Pipe 7	500	50

School

6. 5 LEARNING HOW TO USE EPANET

This example has been relatively easy; using Epanet properly requires a little more effort: a little, but not much. In courses, students generally manage to get to grips with Epanet for designing gravity flow projects in around 20 hours. If you want to get to know the program better, these two books will help you:

- Arnalich, S. (2007). *Epanet and Cooperation: An Introduction to Computerised Water Distribution Modelling,* 200 pages. ISBN 978-84-611-9322-6.

- Arnalich, S. (2007). *Epanet and Cooperation: 44 Progressive exercises explained step-by-step.* 216 pages. ISBN 978-84-612-1286-6.

You can consult them here:
www.arnalich.com/en/libros.html

Arnalich also periodically organises attending and online courses:
www.arnalich.com/en/aula.html

7. Restraining the force of the water

7. 1 INTRODUCTION

A system is nothing more than a container holding many tons of water in movement. The forces generated can be huge and potentially destructive. All your hard work to deliver water to a community can be wiped out in seconds. For this reason it´s vital to understand the forces at work and how to avoid them or restrain them.

7. 2 HYDROSTATIC PRESSURE

The water in a system tends to maintain a state of motion. Any change of direction or velocity generates a force. The basic cases, where the forces generated are shown with a continuous line and those needed to balance them, are:

End pipe. The end of the pipe must contain the pressure of the water.

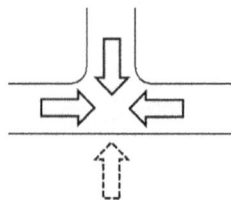

T. The forces from each side left and right cancel each other out. Those from above are unbalanced and produce a net force in a perpendicular direction.

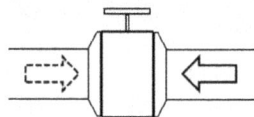

Valves. Shut-off produces a situation similar to that of the end pipe. Partial closing produces acceleration in the direction of the flow. The current pushes the valve in the direction of the flow.

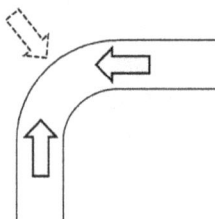

Elbow. This is a situation similar to that of the T, in which none of the forces balance each other. An oblique force is needed to counter-act the two acting forces.

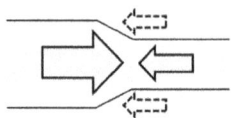

Reduction. In the absence of movement, the pressure from the larger side acts over a greater surface and generates a larger force. If there is flow, acceleration is produced similar to that seen with a valve.

Observe that the pressure tends to push apart the various pipes which are joined to make the pipeline. If over one section there is enough movement for the pipes to separate, the resulting leak is monumental.

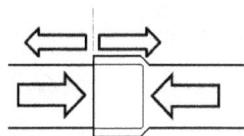

These forces are highly destructive for the installation and must be compensated for.

7. 3 RETAINING BLOCKS

If the burial depth is adequate (more than 1m), the pressure of the terrain is sufficient to hold the pipes together in place. In other situations, however, **retaining blocks** must be used.

A retaining block is a block of reinforced concrete that transfers the forces from the pipe to the terrain. For this to take place, they must be supported by compacted terrain and have a sufficient surface area. The system balances itself, and the pipes are not required to absorb the resulting forces. These blocks are made with a concrete mix of 1:2:4 proportions, and are reinforced to avoid cracking during drying.

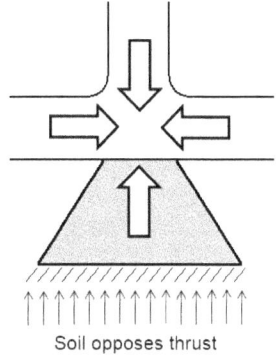

Soil opposes thrust

In the photo, a retaining block is shown for a 90° elbow. The pipe in the middle is simply supporting the mould, and is not a part of the pipe work.

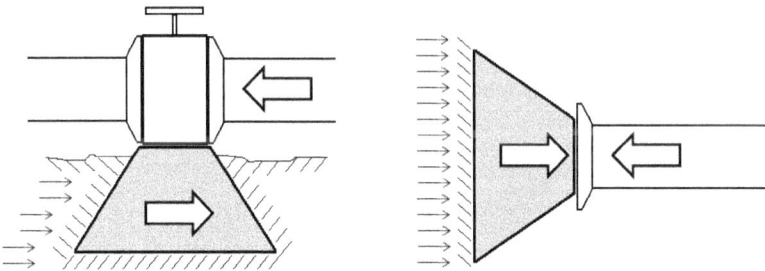

If a concrete valve box is built, this can be used as an anchor.

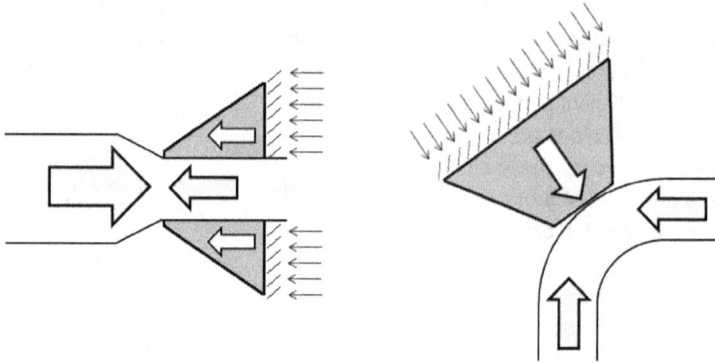

The surface area of the block is determined by the diameter of the pipe, the consistency of the terrain, the internal pressure and the type of disposition to be compensated. Use the following tables to determine the necessary size (Width*Height*Volume):

		High resistance soils: Fragmented rock, gravel…				
ID	PN	Elbow 11° 1/4	Elbow 22° 1/2	Elbow 45°	Elbow 90°	End pipe &T
80	10	0.10x0.18/0.01	0.17x0.18/0.02	0.21x0.28/0.04	0.38x0.28/0.06	0.28x0.28/0.05
	16	0.13x0.18/0.01	0.18x0.28/0.03	0.33x0.28/0.05	0.59x0.28/0.11	0.43x0.28/0.07
100	10	0.11x0.20/0.01	0.21x0.20/0.02	0.29x0.30/0.06	0.51x0.30/0.10	0.37x0.30/0.07
	16	0.17x0.20/0.02	0.24x0.30/0.04	0.45x0.30/0.08	0.77x0.30/0.20	0.57x0.30/0.11
125	10	0.14x0.22/0.02	0.20x0.32/0.04	0.38x0.32/0.08	0.67x0.32/0.17	0.49x0.32/0.11
	16	0.23x0.22/0.03	0.32x0.32/0.07	0.59x0.32/0.14	1.01x0.32/0.37	0.75x0.32/0.20
150	10	0.18x0.25/0.03	0.26x0.35/0.06	0.48x0.35/0.12	0.83x0.35/0.27	0.61x0.35/0.16
	16	0.28x0.25/0.04	0.40x0.35/0.09	0.73x0.35/0.21	1.04x0.45/0.54	0.93x0.35/0.34
200	10	0.24x0.30/0.05	0.37x0.40/0.12	0.68x0.40/0.24	0.98x0.50/0.54	0.86x0.40/0.33
	16	0.30x0.40/0.09	0.56x0.40/0.19	0.87x0.50/0.42	1.46x0.50/1.17	1.09x0.50/0.66
250	10	0.31x0.35/0.08	0.48x0.45/0.20	0.75x0.55/0.35	1.28x0.55/0.99	0.95x0.55/0.55
	16	0.39x0.45/0.16	0.73x0.45/0.32	1.13x0.55/0.78	1.67x0.65/2.00	1.41x0.55/1.21
300	10	0.37x0.40/0.12	0.59x0.50/0.28	0.93x0.60/0.58	1.41x0.70/1.53	1.17x0.60/0.91
	16	0.48x0.50/0.24	0.78x0.60/0.41	1.39x0.60/1.27	2.04x0.70/3.22	1.56x0.70/1.87
350	10	0.43x0.45/0.18	0.61x0.65/0.27	1.11x0.65/0.88	1.67x0.75/2.30	1.26x0.75/1.31
	16	0.57x0.55/0.35	0.93x0.65/0.62	1.49x0.75/1.83	2.23x0.85/4.66	1.84x0.75/2.80
400	10	0.49x0.50/0.25	0.71x0.70/0.39	1.17x0.80/1.20	1.79x0.90/3.18	1.46x0.80/1.87
	16	0.65x0.60/0.49	1.07x0.70/0.89	1.60x0.90/2.54	2.42x1.00/6.45	1.97x0.90/3.86

Source: Saint Gobain–PAM Canalisation (16)

ID	PN	Elbow 11° 1/4	Elbow 22° 1/2	Elbow 45°	Elbow 90°	End pipe &T
		Moderately resistant soils: Sand, clay…				
80	10	0.13x0.18/0.01	0.17x0.28/0.02	0.32x0.28/0.04	0.56x0.28/0.10	0.41x0.28/0.06
	16	0.14x0.28/0.02	0.26x0.28/0.04	0.49x0.28/0.08	0.85x0.28/0.23	0.63x0.28/0.13
100	10	0.17x0.20/0.02	0.23x0.30/0.04	0.43x0.30/0.07	0.74x0.30/0.19	0.54x0.30/0.10
	16	0.18x0.30/0.03	0.35x0.30/0.05	0.65x0.30/0.15	1.11x0.30/0.41	0.83x0.30/0.23
125	10	0.22x0.22/0.03	0.30x0.32/0.06	0.56x0.32/0.12	0.97x0.32/0.34	0.72x0.32/0.19
	16	0.25x0.32/0.04	0.47x0.32/0.08	0.85x0.32/0.27	1.18x0.42/0.65	1.07x0.32/0.42
150	10	0.26x0.25/0.04	0.38x0.35/0.08	0.70x0.35/0.19	0.99x0.45/0.49	0.89x0.35/0.31
	16	0.31x0.35/0.06	0.59x0.35/0.14	1.06x0.35/0.43	1.46x0.45/1.06	1.10x0.45/0.60
200	10	0.29x0.40/0.07	0.54x0.40/0.14	0.83x0.50/0.38	1.39x0.50/1.07	1.05x0.50/0.61
	16	0.44x0.40/0.12	0.82x0.40/0.30	1.24x0.50/0.85	1.79x0.60/2.12	1.54x0.50/1.30
250	10	0.37x0.45/0.12	0.70x0.45/0.25	1.08x0.55/0.71	1.60x0.65/1.83	1.35x0.55/1.11
	16	0.57x0.45/0.19	0.91x0.55/0.50	1.42x0.65/1.45	2.10x0.75/3.66	1.76x0.65/2.22
300	10	0.46x0.50/0.19	0.75x0.60/0.37	1.32x0.60/1.16	1.95x0.70/2.94	1.49x0.70/1.71
	16	0.61x0.60/0.25	1.12x0.60/0.83	1.75x0.70/2.36	2.40x0.90/5.71	1.98x0.80/3.46
350	10	0.54x0.55/0.27	0.89x0.65/0.57	1.42x0.75/1.67	2.13x0.85/4.25	1.76x0.75/2.56
	16	0.73x0.65/0.39	1.20x0.75/1.20	1.91x0.85/3.42	2.69x1.05/8.33	2.20x0.95/5.05
400	10	0.62x0.60/0.38	0.94x0.80/0.78	1.53x0.90/2.32	2.31x1.00/5.89	1.89x0.90/3.53
	16	0.85x0.70/0.56	1.39x0.80/1.71	2.08x1.00/4.75	2.85x1.30/11.63	2.41x1.10/7.03

Source: Saint Gobain–PAM Canalisation (16)

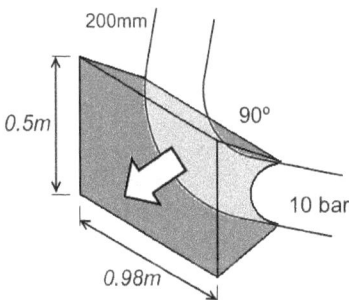

For example, the reading in the tables for a block in highly resistant soils, to retain a 90° elbow of 200mm, with a pressure of 10 bar, is:

0.98x0.50/0.54

This block will measure 0.5m high, 0.98m wide, and will have a volume of 0.54m^3.

Important considerations

- The pressure used is the test pressure (see section 8.7).
- In cases of intermediate diameters, use the dimensions for the corresponding block immediately above.

- For T's, the diameter is that of the perpendicular arm. For a T of 100mm/150mm at 10 bar in gravel, the perpendicular arm has 150mm, and the table recommends 0.61 x 0.35 /0.16, i.e. a block of 0.16m³, 0.61m wide and 0.35m high.
- Reductions are calculated subtracting the corresponding areas from the end pipes. For height the lower value is used, and the width is calculated dividing the resulting surface area by the height. For a reduction of 150mm to 100mm in sand:

$$0.89m \times 0.35m = 0.31 \ m^2$$
$$0.54m \times 0.3m = 0.16 \ m^2$$
$$(0.31 \ m^2 - 0.16 \ m^2)/0.3m = 0.49m \ wide$$

Retention by gravity

In the case of an elbow at the top of a slope, there's no terrain to provide support. To retain the pipe, a block is used that hangs from the pipe. The weight of the block must be greater than the vertical component of the force:

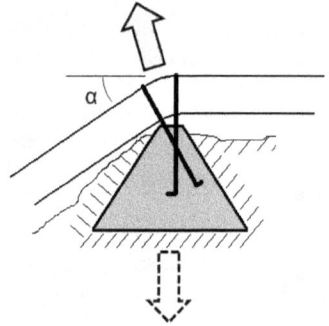

$$V = \frac{10.000 * C_s P * A * sen\alpha}{\gamma}$$

V, Volume of the block in m³.
C$_s$, Safety coefficient, 1.5.
P, Pressure in Bar.
A, Area in m² (A=3.14*d²/4) , d= diameter in m.
γ , specific weight of the concrete in kg/m³, approximately 2400Kg/m³.

With the recommended values, the formula becomes: $V = 6,25 * P * A * sen\alpha$

For the horizontal component, make sure the surface area of the block on the slope side is greater than A*(1-cos α), where A is the product of the width by the length of the corresponding elbow in the tables above.

Calculation example:

Calculate a gravity block for a 200mm ID pipe, with an internal pressure of 10 bar and a 30° angle, in fractured rock.

Vertical component:

$V = 6.25 * P * A * \text{sine } \alpha$; $A = 3.14*d^2/4 = 3.14* 0.2^2/4 = 0.0314 \text{ m}^2$

$V= 6.25 * 10 * 0.0314 * 0.5 = 0.98 \text{ m}^3$

Horizontal component:

As there is no value for 30°, the angle immediately above is used, 45°. The reading is 0.83x0.50/0.38.

$A = \text{width } * \text{ height} = 0.83 * 0.5 = 0.415$

$S > A*(1\text{-cos } \alpha)$ → $S > 0.415 (1 - 0.866)$ → $S > 0.055 \text{ m}^2$.

For the retaining block to be a protective element, it´s very important that the pipe is supported by it over the maximum possible area. Otherwise, the forces are concentrated on one point and the effect becomes counterproductive. For the sake of clarity, the drawings show specific points of support, although in practice they should provide ample support for the pipe:

Insufficient support
(Forces are concentrated)

Correct support
(Forces are dispersed)

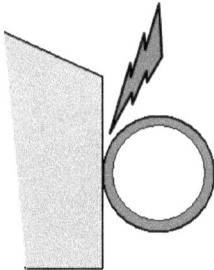
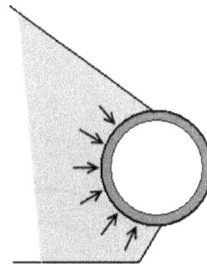

Pipe section

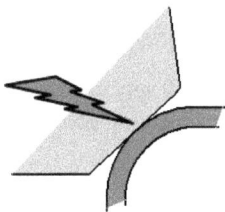
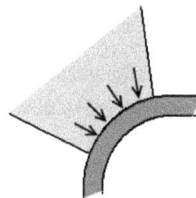

Top view

7. 4 WATER HAMMER

In the days of castles and sieges, the battering ram was used to destroy fortifications and castle doors. Inside the ram, soldiers would pull it back and then let it smash into the door of the fortress.

If the sudden pound of the ram hitting the door was enough to inflict major damage on the fortress , imagine what could happen in a water system if a similar effect took place, where various kilometres of a column of water were suddenly stopped dead inside a pipe.

Avoiding damage

The water hammer effect takes place when accessories are installed which cut the flow off very abruptly (automatic shut-off valves, ball valves, simple air valves) or due to the presence of air inside the pipes. The starting and stopping of pumps also generates large water hammers, although we won´t deal with them here, as we´re looking only at gravity flow distribution.

To avoid the water hammer damaging the installation:

1. Try and avoid the pipeline taking a route in which there are high points between two descents (a drop of only one meter allows air to accumulate). Place air valves only where you have to. In some points you can install a public tap stand instead of an air valve, if the demand dictates it. The air will be expelled from the pipe when the users open the tap.

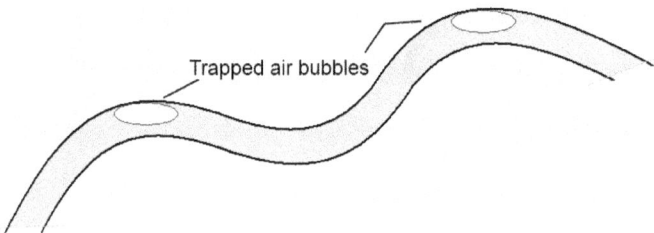

Trapped air bubbles

2. Don´t install valves which an operator can close off very quickly. Instead of ball valves (left), use gate valves (right), for pipes of over 50 mm.

Abrupt shut-off

Gradual shut-off

3. Avoid accessories which rapidly change the velocity of the water, automatic shut-off valves, check valves or air valves.

4. Make sure you´re always designing for the pipes to be full. Emptying and filling pipes leaves lots of residual air which is then difficult to get rid of.

5. Divide up long reaches of pipe so you reduce the amount of water which could be affected by a water hammer. This will also help you work out in which reaches there is a blockage.

6. Do the calculations to avoid surprises.

Calculating the Water Hammer

What you need to know is the excess pressure the pipe will be dealing with, to see if it´s within the specified limits, together with the force exerted on the supporting structures for the pipe.

First of all, the speed of the pipe, *a*, for water, the formula is:

$$a = \sqrt{\dfrac{1}{1000 * (\dfrac{1}{\varepsilon} + \dfrac{D}{Eg})}}$$

Where: a : speed in m/s
ε : Unit of elasticity of water (2.05 * 10^9 Pa)
D : Internal diameter in meters[3]
g : Thickness of the pipe in meters[4]
E : Unit of elasticity of the pipe: PVC = 3.3 * 10^9 Pa
HG = 207 * 10^9 Pa
HDPE = 0.8 * 10^9 Pa

---

[3] See the friction loss tables in Appendix B.

If the <u>shut-off is abrupt</u>, the Allievi formula is used: $\Delta H = a * \dfrac{\Delta V}{g}$

Where: ΔV : Variation in velocity, before and after the hammer.
g : Gravity (9.81 m/s^2).

If the <u>shut-off is gradual</u>, the Michaud formula is used: $\Delta H = \dfrac{2L\Delta V}{gt}$

The shut-off is abrupt if the stopping time t meets: $t < \dfrac{2L}{a}$

To get an idea of the force exerted on the installation, you can use this formula:

$$F = P * A$$ P, pressure in bar ; A, area in cm^2 ; F, force in Kp.

Calculation example:

Calculate the water hammer and the force the valve will be under in an HDPE PN 10 bar 160 mm pipe, if the reach is 200m long, the pipe carries 15.64 l/s, the pressure the valve is put under is 5 bar, and the shut off lasts 1 second and 5 seconds respectively.

$$a = \sqrt{\dfrac{1}{1000 * (\dfrac{1}{\varepsilon} + \dfrac{D}{Eg})}}$$

$$= \sqrt{\dfrac{1}{1000 * (\dfrac{1}{2.05 * 10^9} + \dfrac{0.141}{0.8 * 10^9 * (0.16 - 0.141)/2})}} = 229.7 m/s$$

The time beyond which a shut-off is considered abrupt is:

$$t < \dfrac{2L}{a} = \dfrac{2 * 200}{229.7} = 1.76s$$

For 1 second, the <u>shut-off is abrupt</u>, and the Allievi formula is used:

From the friction loss tables, at 15.64 l/s the velocity is 1 m/s:

$$\Delta H = a * \dfrac{\Delta V}{g} \quad \Delta H = 229.7 m/s * \dfrac{(1m/s - 0)}{9.81m/s^2} \quad \Delta H = 22.97m = 2.3bar$$

The maximum registered pressure is 2.3 bar + 5 bar = <u>7.3 bar</u>. The force the structure will be subjected to is:

A = 3.14*d^2/4 = 3.14 * 14.1^2/4 = 156cm^2 F = P * A = 7.3 bar * 156cm^2 = <u>1140 Kp</u>

For 5s, the <u>shut-off is considered gradual</u>; the Michaud formula is used:

$$\Delta H = \frac{2L\Delta V}{gt} = \frac{2*200m*1m/s}{9.81m/s^2*5s} = 8.15m$$

F= (0.815+5)bar*156cm^2 = <u>907 Kp</u>

One Kilopond (Kp) is the force with which standard gravity attracts a mass of one kilogram. To visualize the force, think of 10 Kp as what you would need to support 10 kilos.

7. 5 ANCHORS ON SLOPES

The steeper the slope, the more the weight of the pipe, and the water in it tends to push the pipe downhill. As a general norm, anchors are not needed if the slope is less than 25% for buried pipes, and 20% for pipes above ground.

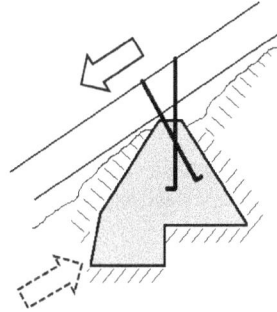

Solid rock

If it´s impossible to excavate, the pipe must be on top of the rock. As plastic pipes are more fragile, and as PVC disintegrates over time in sunlight, galvanized iron pipe must be used. If the climate is very cold and there is a risk that the pipe may freeze over night, the pipe carrying the water is placed inside another larger pipe. If the internal pipe is plastic it will be better insulated.

The rock is perforated and expansive hooks are put in place. Supports are welded onto the pipe and encased in concrete. A support like this is required for every pipe length, to provide sufficient support, and to prevent damage to the threaded unions due to flexing:

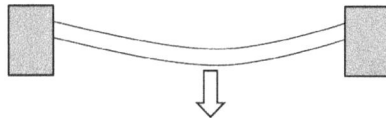

Other soils

If it's possible to excavate, the most simple solution is to get down to the bedrock and follow the guidelines above, without the need to place supports for each pipe length, as these will be supported by the trench floor.

If it's not possible to excavate and there are no alternative routes, you can use this rule of thumb, used for hydroelectric systems:

One block every 30m, encasing the pipe in 1m^3 of concrete for every 300mm of diameter. If there is an elbow of less than 45°, the volume is doubled. If the elbow is 45° or more, it is tripled.

This rule is valid for relatively small pipes, which you'll use in the large majority of cases, and drops of less than 60m.

Calculation example:

The route of a pipeline, of 200m, over a slope of 36%, is shown in the diagram. Work out where the anchors are required:

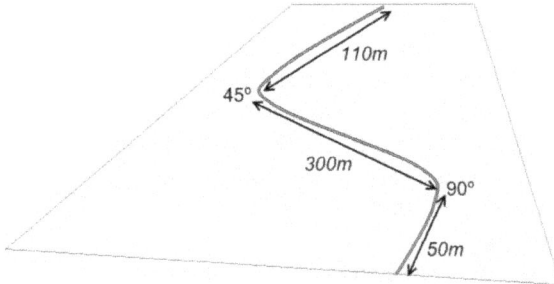

At the start of the slope, a gravity anchor is needed. The anchors for the slope will be:

200mm * 1m^3 of concrete/300mm pipe = 0.67m^3 for each normal anchor.

The 45° elbow requires double the volume: 0.67m^3* 2 = 1.33 m^3

The 90° elbow requires triple the volume: 0.67m^3* 3 = 2 m^3

Following the length of the pipe, the required anchors are:

Cum. length	Volumen	Cum. length	Volume
30m	$0.67m^3$	260m	$0.67m^3$
60m	$0.67m^3$	290m	$0.67m^3$
90m	$0.67m^3$	320m	$0.67m^3$
110m	$1.33m^3$	350m	$0.67m^3$
140m	$0.67m^3$	380m	$0.67m^3$
170m	0.67m	410m	$2.00m^3$
200m	$0.67m^3$	440m	$0.67m^3$
230m	$0.67m^3$		

8. Pipe installation

8. 1 INTRODUCTION

The correct installation of pipes has an enormous impact on the duration of the installation, maintenance requirements, leaks, and water quality. In areas where the pipe is damaged, liquid is sucked into the pipe from the surrounding area as it empties, drawing in all kinds of muck. At points where accessories are damaged, suction will take place due to the Venturi effect:

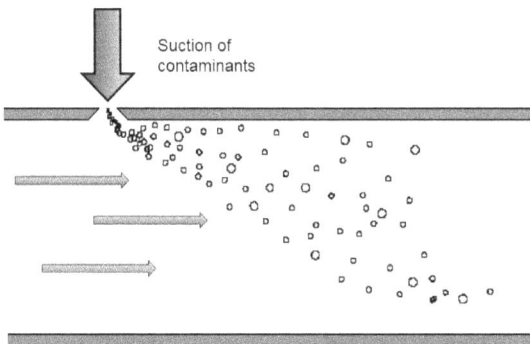

Suction of contaminants

8. 2   CEREMONIES

The ceremonies at the start of projects and on completion are one of the most valuable tools you have at your disposal. In the first place, because they imbue a sense of importance of the job at hand in the eyes of the end users, and this favours well for future care and maintenance. But they also have other interesting effects amongst the workers: they break the ice and they create a sense of teamwork and transcendence for taking part in an important project.

Fig 8.2. Ceremony for the project opening, Proja Jadid, Afghanistan.

If you think they are a boring commitment, you´re missing out on a great occasion!

8. 3   EXCAVATION

The pipes are buried to protect them mechanically, but also to protect them from interference and from the cold. The depth, measured from the top of the pipe, depends on the case at hand:

1.   In normal conditions 1m.
2.   Under frequent traffic, 1.5m.
3.   In cold areas, below the soil freezing level in winter.
4.   In areas unlikely to be inhabited, without risk of freezing, and with no traffic etc., you can save with a shallower burial of 0.5m.

The depth of the trench for excavation is:

h = Depth of burial + Pipe diameter + 10 cm

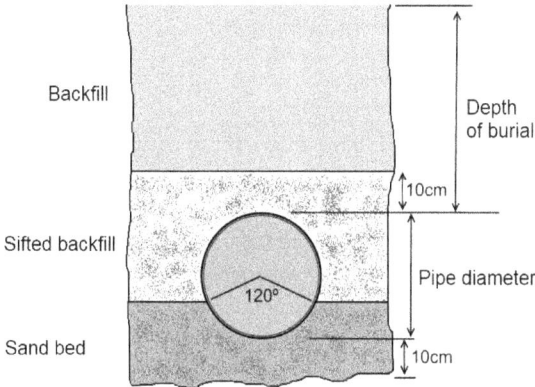

The width of the trench is the minimum required to be able to work. It will depend on the cohesiveness of the terrain, the condition of the workers, the depth and the machinery used. In normal conditions, 80cm tends to be a good compromise between comfort and economy.

Pay special attention that the slope in the trench has no abrupt unintentional changes in depth. A way of checking for this is to cut 3 bars to the same length, one which can be comfortably used by the person who is going to use it. If the slope is more or less continuous with gradual changes, the 3 bars will remain more or less aligned. If there is a sudden change, one will stand up from the others:

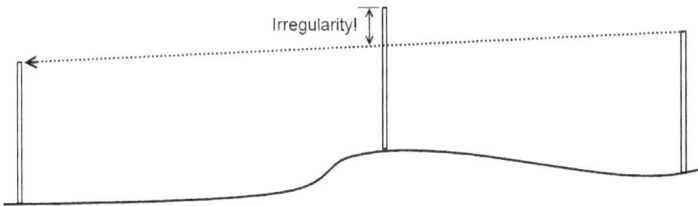

Precautions:

1. **Protect the soil that has been dug out** at all cost. In many areas, loose soil is a coveted good, and the temptation to not have to dig to get hold of it is great. If you´re not careful, you´ll end up having to buy back your own soil without having budgeted for it.

2. In **deep trenches or unstable soils** the risk of accidents due to sudden collapse is great. To be safe, the trenches can be dug in a **V shape**.

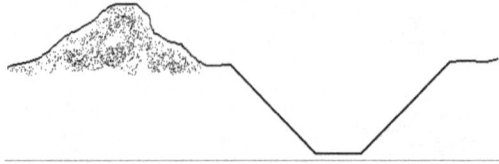

3. **Don´t start the trench digging too soon.** If you do, you´ll not only inconvenience the locals, but they will also fill with rubbish, get shallower and can flood. There is also a tendency to confuse trenches with waste water channels. Wait until you have the pipes before starting with the excavation. Delays in material deliveries can leave trenches open for months. Ideally, the digging team and the installation team should work in coordination, so that at any given time there are only a few hundred meters of trench open.

4. In areas with little social cohesion, in the process of being inhabited or with a disorderly population, **keep the route a secret** and above all don´t mark out the route ahead of time. In many places, the news of an installation is a call to build close to the future pipe location. Then, when you come to install the pipe, the route could be obstructed by endless houses.

5. **Keep an eye periodically on the route** so you can negotiate quickly in relation to any building in the area. Demolishing partially finished buildings in which people have invested what little capital they have, is problematic.

6. During the excavation, have ready at hand **a reserve supply of repair materials** including pipe, electrical and telephone cable, and any other things that may get damaged, to re-establish the service as quickly as possible.

7. If the route passes through traffic areas, have metal plates at hand so as not to stop the traffic.

Economic injection potential

Earth moving is a job that requires a lot of labour. It can be a unique opportunity to inject money into an area that is economically depressed or with high unemployment. Every meter of trench corresponds to around 1 day of labour. Moreover, digging with machinery is often more expensive and can cause severe delays in areas where spare parts are not readily available.

Two important precautions before embarking on a project of this kind are paying suitable wages which don´t compete unfavourably with other activities (agriculture, education...) and which don´t take place at times when the community needs a lot of manpower (e.g. during harvest time).

8. 4   PIPE INSTALLATION

In the community there tend to be people who have experience in pipe installation. Here we´ll just be dealing with the installation of HDPE pipe and the most common mistakes:

1. **Ignoring thermal contraction**. Pipe tends to be installed when they are dilated due to temperature. If they are installed without special care, at night time they will contract, reducing the overlap distances and even pulling pipe sections apart. With threaded or HDPE pipe which can´t be disconnected, stresses are created which will damage other parts of the installation.

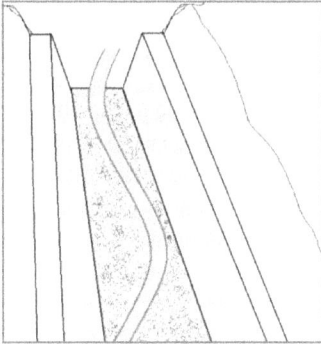

To avoid this, pipes should be installed in the cooler hours of the day, and laid in "S" shapes. One method, which is uncommon as it involves an extra burden on the workers, is to install pipe in the morning, dig trench at the hottest time of day, and continue installing pipe in the afternoon.

2. **Exposing PVC pipes to the sun or heat.** The result is that they deform. Installing banana-shaped pipes is exasperating, often requiring the trench to be widened.

3. **Using toxic oils or grease as lubricant**. Use comestible oils (maize, palm oil...) and above all avoid toxic ones (engine oil, mechanical grease...).

The process of Butt fusion welding HDPE can be seen in appendix F, due to its interest as a choice of material and for the relative lack of knowledge on the part of technicians.

8. 5 BACKFILL

This is a delicate operation. The majority of early breakages or accidents occur because these instructions are not followed.

First of all, remove any protruding material or sharp object from the trench before laying the **sand bed**.

Think of the sand bed as packaging for a very delicate piece of equipment. Ensure there is an even and smooth surface on which the pipe can be laid. A pipe full of water resting on a stone will quickly break, as all the pressure becomes concentrated on one point.

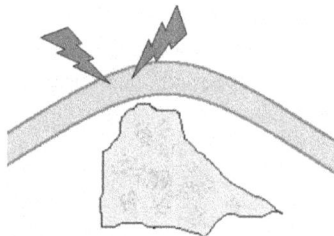

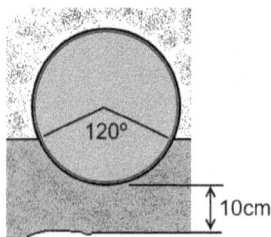

The minimum thickness for the sand bed is 10cm. Above this, and depending on the pipe diameter, you should refill until you have a guaranteed angle of contact of at least 120°.

Then further backfill is added to 10cm above the pipe with **sifted excavated material**, again to avoid damage to the pipe. The sifted material can be passed through a wire mesh mounted on a wooden frame, as in the photo.

After this, a further **15cm** layer of the **original material** dug out from the trench is added and compacted. A simple way of compacting in the absence of machinery is to use a cylindrical drum filled with 100 litres of water.

Over this is placed plastic **signalling tape**, warning people of the pipe in case of future construction works in the area. If tape without legible warnings on it is not available, use cordoning tape. Sticky tape is too common and fragile to draw sufficient attention.

Finally, the remaining trench is filled and compacted in layers of 15cm until completely refilled.

With PVC, galvanized iron pipe and threaded unions, the joints are left unburied for visual inspection during the first pressurised system test.

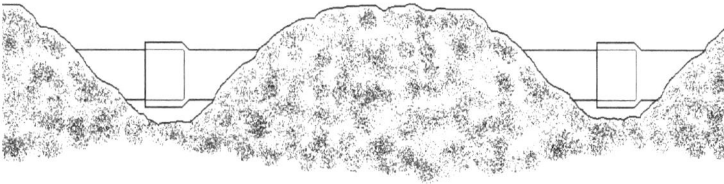

Note in the photo that the sand bed is inadequate. Sand is quickly lost in gravel, which is why it is not recommended when choosing a route. If there is no other option, use iron pipe, which is much more resistant.

Finally, **signal the pipe with markers** where there are no other clear points of reference: open fields, cultivated lands, or steep slopes...On the markers, write the pipe diameter, material and a small diagram if there is an elbow or "T". The idea is to be able to find the pipe quickly for repairs or system enlargements.

Metal pipes and corrosion

Often small metal galvanized pipes are used in highly corrosive soils. In these kinds of soil plastic pipe is the material to use. Note in the photo how corrosion has broken the saddle bust for a domestic connection, only 5 years after initial installation.

If soil tests are not available, the most problematic areas are:
- Low points or dips, which are often more humid
- Water courses, damp areas
- Puddles, swamps, lakes, peaty areas and areas rich in organic acids, bacteria etc.
- Estuaries, marshes, saline soils near the sea.

When soil pH is less than 5.5, it is potentially highly corrosive. Clay soils are less of a risk, and gypsum and organic soils are highly corrosive.

Never install metal pipe in contact with other metals. The presence of a different type of metal can cause currents to run between them and lead to rapid galvanic corrosion.

If metal pipes can´t be substituted for whatever reason, they can be protected in polyethylene sleeves, with sacrificial anodes or cathodic protection. Don´t underestimate corrosion. Up to 5% of the gross domestic product of a country can go towards fighting corrosion. If you remember the famous 0.7% campaign, you´ll realise that in many places this kind of money won´t be available.

8. 6 TROUBLESOME SITUATIONS

In some areas special care must be taken when installing pipe:

Under heavy traffic

On the one hand, the weight of the vehicles can break the pipe, and on the other hand a leak could cause an earth collapse. To avoid this, bury pipe at 1.5m depth, inside a larger metal pipe, which extends 5m either side of the road. The exterior pipe protects the one inside from the weight of passing vehicles, and expels any leaks to either end. Some animals, like elephants, buffalo, giraffes etc., can cause damages similar to that caused by vehicles.

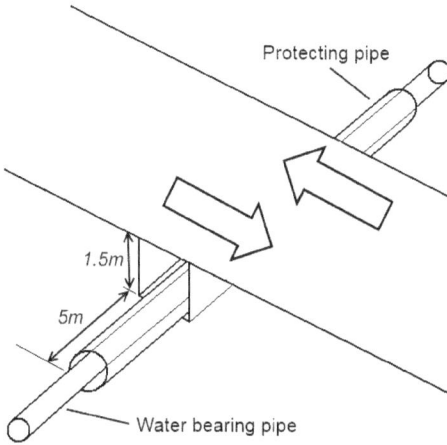

Protecting pipe

1.5m

5m

Water bearing pipe

Insufficient depth

In some places, the bedrock, the water table, or other obstacles, impede sufficient depth beneath traffic or heavy animals. In these situations, the pipe is protected by pouring a 10cm thick concrete layer to form a slab laying over undisturbed soil. The distance between the slab and the pipe should be sufficient for it to be able to break without damaging the pipe.

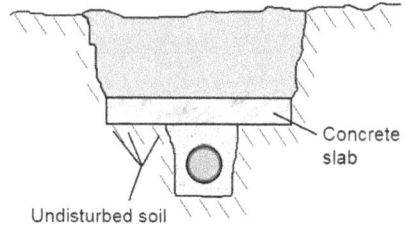

Concrete slab

Undisturbed soil

Suspended crossings

Surge channels and narrow streams can be crossed with one suspended pipe length of galvanized iron pipe. There can be no unions, so with 1 meter either side for pipe support, the maximum possible distance is 4m. If there is a risk of erosion install gabions.

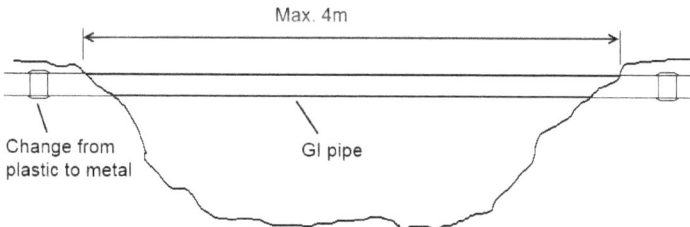

Max. 4m

Change from plastic to metal

GI pipe

Larger crossings require support in the centre or tensile cable. Note how in this bridge, 2 pipes are used without support, leaving a bolted union in the middle. Despite the union being stronger than a threaded one, the result is that the pipe has drooped significantly and the entire installation is under stress:

One way of making a support in shallow water is with a dambo pillar. This consists of a square slab 15cm thick, and 1.3m on either side, on top of which an oil barrel has been placed at the moment of pouring. On top of this a second drum has been welded. A 20cm layer of stones is laid first as a foundation. The slab rests on top of this, spreading the weight evenly. The barrels allow for cement to be poured in without interference from the surrounding water.

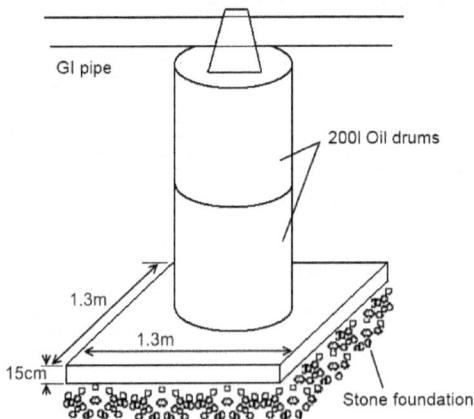

Rivers

If there´s a bridge, the simplest is to run the pipe along it with brackets, as in the photo. The brackets need to be at least every 6m. All exposed pipe must be galvanised iron. Plastic pipes degrade and deteriorate before long. Look at how they have come apart in the photo:

If there is no bridge, you can choose between a suspended crossing with a cable, or a submerged crossing. The details regarding suspended cable crossings can be found in appendix E, of reference 12 of bibliography.

For submerged crossings, use HDPE. An 8mm steel cable can be used as a guide. Every 6m, a ballast is fixed to the pipe with a flange. Each ballast is compensated with a buoyancy float. As the cable is pulled from the opposite river bank the pipe is gradually introduced into the river. Once the pipe is in place, the buoyancy floats are removed and the ballasts drop to the bottom.

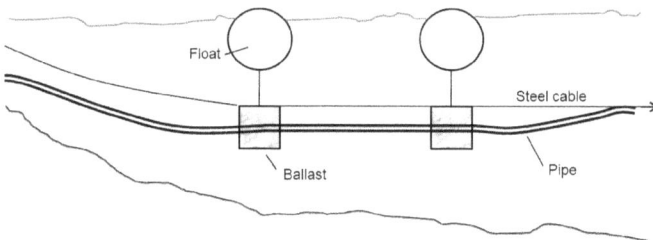

If 200L oil drums are used as buoyancy floats, a 40cm x 40cm concrete ballast is sufficient.

Contaminated areas

Latrine areas or rubbish dumps are very dangerous due to the potential for liquid suction. Install HDPE in these areas. If this is not possible, protect the unions by sealing them in polyethylene sleeves or with blocks of clay.

8. 7 PRESSURE TEST

A pipe section pressure test is one in which the pipes are subjected to pressures slightly higher than normal working pressures for a specific length of time. The goal is to find defects in the installation and leaks.

The test process is meticulously described in terms of time, pressure, lengths…However, in the field it´s unlikely you´ll be able to work with this much precision, due to a lack of materials, poor material quality and an absence of control conditions. If you try to do a rigorous test you end up wasting time checking that all the measuring equipment is properly calibrated, whether the pipe has dilated in the sun or if something is broken.

This test is extremely valuable to identify defects, so even if you´re not measuring hundredths of a bar at exactly 10ºC, use this simplified method:

1. Select pipeline sections of less than 500m.

2. With all the retaining blocks installed correctly and the pipe partially buried, seal one end of the pipeline and fill it with water, allowing the air to escape. The concrete in the retaining blocks should have had at least 7 days to harden sufficiently.

3. Seal the other end with the special cap that comes with the pressurising equipment. Increase the pressure up to 1.5 times the working pressure, or up to 80% of the nominal pipe pressure limit. Maintain the pressure for at least an hour.

4. Do a visual inspection of the joints and unions to check for leaks.

Avoid the temptation of doing without the pressure test. There are few things that frustrate project personnel more than a system which looses all its water before it reaches its destination, having worked for months to prepare the project. The test will warn you of basic human errors like installing PVC unions the wrong way round, before they sabotage the whole project.

8. 8 POST-CONSTRUCTION DISINFECTION

During the installation, the pipes have been in contact with the ground, site materials and waste, and surplus oil or soap used to join the unions. After a new installation or repair the system has to be disinfected. The process is as follows:

1. Let the water run through the system for several hours to clean out organic materials.

2. Feed a solution of at least 50mg/l of chlorine into the system, until you´re able to measure the same concentration at the point furthest away.

3. Leave the water in the system for at least 3 hours. Once that time is up, check that there is still a sufficient chlorine concentration (over 25 mg/l).

4. Let the water run through the system until all the excess chlorine has been eliminated.

For water, 1mg/l and 1 ppm (part per million) is the same thing. To prepare these solutions you can use granulated chlorine (calcium hypochlorite), or bleach, so long as it's apt for water disinfection.

Preparing the solutions

If you´re using granulated chlorine (calcium hypochlorite), the concentration is approximately 70%. That means 1 kg of the product contains 0.7 kg of chlorine. To obtain a solution of 100mg/l, you need:

$$X = 100/0.7 = 0.143 \text{ g/litre of solution}$$

If you disinfect from a tank, it´s simple. For example, if the tank holds 100m^3, you need to mix:

$$100{,}000 \text{ l} * 0.143\text{g/l} = 14{,}300 \text{ g} \text{ or } 14.3\text{kg of the product.}$$

For practical purposes, it´s simpler to dissolve these 14.3kg in a container of a 100 litres and pour it into the tank, rather than trying to mix it directly in the tank itself. A soup spoon holds approximately 14gr, i.e. 10gr of real chlorine if the product is calcium hypochlorite.

The concentration of chlorine in bleach is very variable, between 4% and 35%. Assuming it is 25% and in the case of the tank:

$$X = 100/0.25 = 0.400 \text{ g/litre of solution}$$
$$100{,}000 \text{ l} * 0.400\text{g/l} = 40{,}000 \text{ g or } 40 \text{ kg of the product}   \text{(approx. 40 litres)}$$

On millilitre contains approximately 20 drops.

Precautions

Chlorine is an irritant. It should be handled with adequate protection. Some of the solutions needed for disinfection are highly concentrated. Don´t store chlorine in bottles which can be confused with drinks.

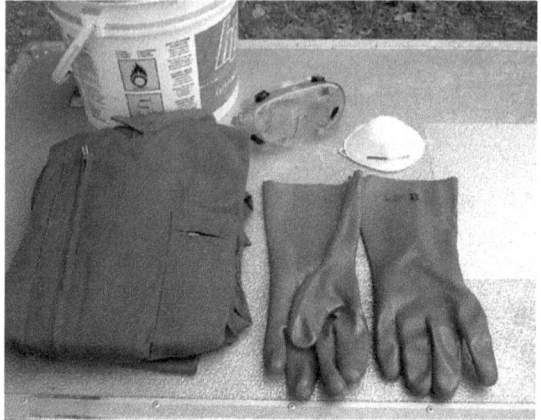

Chlorine is a powerful oxidant. If it´s stored with other equipment, especially metals, considerable damage can result. Look at the damage to a portable treatment plant, in which a container with 5 kg of chlorine was stored for a week. Chlorine storage should be in a separate place.

9. Piping and fittings

9. 1 INTRODUCTION

Water systems are made with materials. Their selection becomes a key factor in the day to day working of a project. Some general principles, debatable in some cases, are:

- **Use materials which are accessible.** Not necessarily local, but accessible.

- **Use well known materials and technologies** locally or introduced with adequate training. Sometimes a project is a unique opportunity to introduce superior technology. Often, introducing a technology in the classroom leads to disaster as it gives the individual concerned a false sense of choice and encourages them to use it without becoming sufficiently familiarised with the technology first. Most adults don´t appreciate sitting down in a classroom "like children" in a subordinate role to the teacher. On site however, that person is being paid to apply a technique as a recognised professional, allowing him or her to develop an interest in it and understand its professional advantages.

- **Don´t use one-off parts which are expensive.** The more expensive a component the less likely that it can be replaced. If many systems fall into disrepair because a new generator or pump can´t be paid for, can you really expect obscure working accessories like pressure reducing valves to be replaced?

One clear exception to some of these rules is the use of HDPE, which, in my opinion, is the material of choice.

Logistics

Material logistics are vital. There are few things that can delay or cause a project to fail more than receiving a delivery of materials which is inadequate. In the case of pipes, this can be key as they incur considerable transport costs.

Quality

There are materials which are, for all intents and purposes, useless. In many places, the person who supplies the materials is not a professional, and may often buy them as if they were plastic urinals or vitamins for camels. To improve the profit margin this person may buy much cheaper materials without understanding the consequences, and you may end up with faulty parts or second hand materials. It becomes very difficult to negotiate a refund if the consequences for the supplier means bankruptcy.

Organise material inspections before making any payment, and ask for samples (i.e. pipe sections) of the material quality you´ve asked for. Inspect the material before it leaves for its destination to avoid having to cover additional transport costs for refunds.

9. 2   PIPING

Piping is the fundamental component of the project. Here only PVC, HDPE (high-density polyethylene) and GI (galvanized iron) are dealt with. Other pipe such as ductile casting, asbestos, and cement, are not mentioned here as they are generally not appropriate in these kinds of contexts.

PVC

This is the most popular pipe due to its low cost, ease of installation and universal availability. It´s produced in almost all diameters. However, it is considered highly toxic due to the releasing of dioxins and heavy metals. Various ecological organisations such as Greenpeace have launched campaigns to replace PVC with other products.

Pros:
- Low cost, often being the cheapest pipe available.
- Universal availability.
- Inert. Does not react with chlorine or with most compound chemicals. Does not corrode.

- Know-how: in most places the know-how and skills for installation are available.
- Light-weight. Cranes are not needed even for the largest diameters. Ideal for areas of difficult access, as pipes can be transported, offloaded and handled by people:

Cons:

- Light sensitive. PVC degrades if exposed to the sun.
- Brittle at low temperatures.
- Frequent maintenance, with low mechanical resistance and unions which could be improved.
- Tendency to break length-wise. In this kind of breakage the pipe opens up end-to-end, leading to massive leaks.
- Rapid reduction in resistance to pressure with temperature. At 43ºC, exposed to the sun, resistance is halved.
- Toxicity in the long term, causing environmental problems.
- Joining to accessories is done with adaptors.

Types of joints:

Socket-spigot. The pipe has one expanded end and one normal. The normal end is covered with special PVC glue and is introduced into the expanded end.

Threaded. This is rare, and usually only in small diameters to transition to another material, usually GI. In large diameters they are used to join staggered sections for a borehole.

Z-joint. This is the typical union for pipes of over 50mm, with a gasket which must be inserted in the right direction so that the water exerts a force on the lip of the union, pushing it against the pipe wall and sealing the joint.

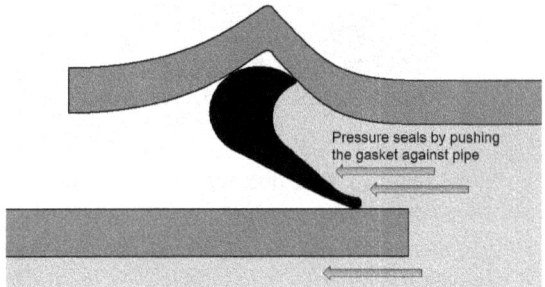

Pressure seals by pushing the gasket against pipe

Flange. This is not a common pipe-to-pipe union. It´s used to transition between PVC and metal, or PVC and an accessory. Attention must be paid to make sure the pilot holes between each part are the same. A gasket is placed in between the bolted section.

With all joints where one pipe is introduced into another, pay special attention to the direction of flow. A path in which the water *walks down steps* generates less friction.

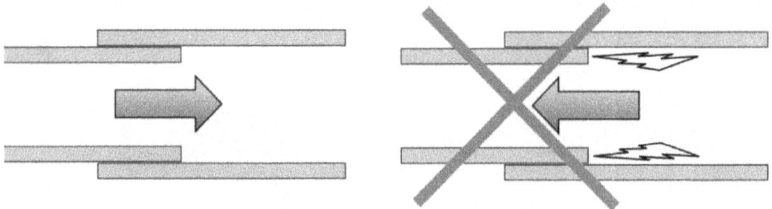

Fig. 9.2. PVC Z-joint. Flow from left to right (→) .

GI pipe

Galvanized iron is the best option when pipes have to be installed above ground. They are popular in small systems and inside buildings.

Fig. 9.2. Bed rock pass with GI. The union is housed inside the block.

Pros:
- Mechanical solidity.
- Stable in sunlight.
- Universally available.
- Direct connection to accessories.
- Can be welded. Although much of the protection against corrosion is lost on welding, accessories, anchors and elbows can be welded at almost any angle.
- Joining with concrete. This is the pipe to use to pass through concrete walls in tanks and other components. In this case, the minimum wall thickness should be 30cm.

Cons:
- Suffers from corrosion.
- More expensive.
- Is not inert. GI pipe consumes chlorine and reacts with water.
- Tends to leak.
- Threaded unions become a little impractical for diameters over 150mm.
- The tools needed for cutting, threading and joining are expensive above certain diameters.

Types of joints:

Threaded. The pipe is threaded on both ends and is joined using threaded ring (sockets).

The join should be made using Teflon tape for smaller diameters and natural fibre material for larger diameters, placed along the threads for water tightness. If the wrenching tools are not used correctly, the pipe is left with teeth marks and scratches.

Flanged. The pipes are joined by bolting the unions together. This is the most resistant union.

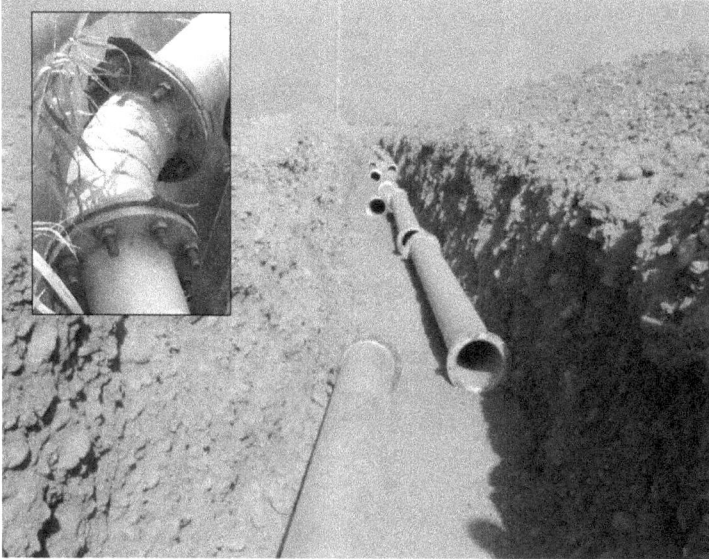

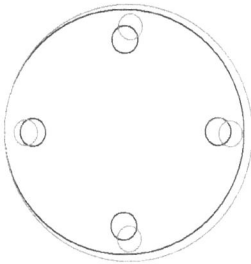

With this kind of union, it´s vital to check that the standards between accessories and pipe are the same. Using two different standards means the holes do not align and joining is impossible. The thickness of the flange and the hardness of the material makes it virtually impossible to drill new holes.

Union. When pipes are threaded systems quickly become impossible to assemble and disassemble. For a T, for example, the whole installation would have to be rotated to assemble the pipes. Moreover, a breakage at one point will require disconnecting the entire installation until that point is reached. The placing of unions is marked with arrows in the photograph.

To allow assembly and disassembly, unions are placed at all critical points. They consist of two housings with external threads, which have a threaded joint at each end to connect to the pipe. Once the pipes are joined to each housing, the external central ring can be tightened and loosened.

HDPE pipe

Polyethylene, is, in my opinion, the material of choice. It´s inert, cheap, mechanically resistant, and the unions provide such a tight seal that it´s even used to transport gas. Moreover, for diameters up to 90mm, the pipe comes in rolls of tens and up to hundreds of meters, which makes installation much simpler and adapts to specific applications such as installing pipe underwater.

Consult Appendix D for details of weight and meters per roll.

Pros:
- Almost total water tightness.
- Available in rolls of several meters.
- Inert and corrosion free.
- Mechanically resistant. The unions are more resistant than the material of the pipe itself. Putting welded pipes through a traction test often results in the pipe breaking before the joint itself.
- Extremely light.
- Low cost. In oil producing countries it can be surprisingly cheap.
- Requires fewer accessories for corners and curves.

Cons:
- A technology which is often unknown, although it can be learnt quickly.
- Slowness in the installation of pipes sizes that don't come in rolls due to the time required to leave the pipe at rest.
- Joining to accessories with adaptors.
- Initial investment in the welding equipment.

Types of joints:

Electro fusion. The accessories have two electrodes. To join them, the current which passes through the accessory heats the HDPE until fusion is achieved. This is a method which is not recommended for cooperation projects due to the cost of the accessories and the high cost and sensitivity of the required equipment.

Butt fusion. For this kind of joint a much cheaper and more robust piece of equipment is used. The most basic models are available from 4000$ upwards. It consists of heating the ends of each pipe until they melt, at which point they are joined under pressure (see Appendix F). The resulting joint has no transition. The material is continuous along the joints, as you can see in the right hand image, which shows a longitudinal section of a joint:

Compression. Allows the joining of accessories without machinery. If the welding equipment is damaged, the local population will have an alternative. This is the quickest pipe-laying technique and the one to use in emergency interventions. However, they tend to leak between accessories. Aside from emergencies, avoid using these if welding equipment is available.

Logistics and pipe quality

To order pipe 5 parameters need to be mentioned: material, diameter, nominal pressure, type of union and presentation. For example:

Galvanized iron pipe, 25mm, 25 bar, threaded union and in 6m lengths.
HDPE pipe, 110mm, 10 bar, Butt fusion welded and in rolls of 50m.

- Due to storage difficulties, manufacturers tend to produce pipe on demand and have very little material in stock. It´s common for there to be a waiting list of several months. Order the pipe as soon as the design work is completed.

- Pipe comes in lengths of exactly 6m. By a fraction, this means you can´t shut the door of a 20-foot container with the pipes in. You will need to store them on-site in a 40-foot container.

- Don´t forget to order the bolts, nuts, and washers you will need for the joints.

Regarding quality:

- Plastic pipes should be of a uniform colour, with no stretch marks or grooves which indicate that the pipe has been extruded cold. To get an idea of what this looks like, fold in half a sheet of plastic or stretch a dark coloured plastic bag.

- Free of superficial defects. With PVC there are often bumps and bubbles.

- Free of shape distortions. Banana shaped pipes are very difficult to install. Pay special attention that the ends are not deformed, making sure they are regularly shaped to allow for joining. Sometimes the ends are elliptical, which makes joining almost impossible.

- Iron pipes should have threads free of damage and rust, with no tool marks or scratches.

- For HDPE pipe in rolls, make sure they have not been bent beyond their maximum curvature. When this happens, a notch is left visible.

9. 3 BASIC FITTINGS

Leaving aside the obvious: elbows, T´s etc.:

Joining accessories between different pipe materials

All have basically the same system. The compression of a gasket between two flanges causes it to expand until it meets the pipe, sealing the joint:

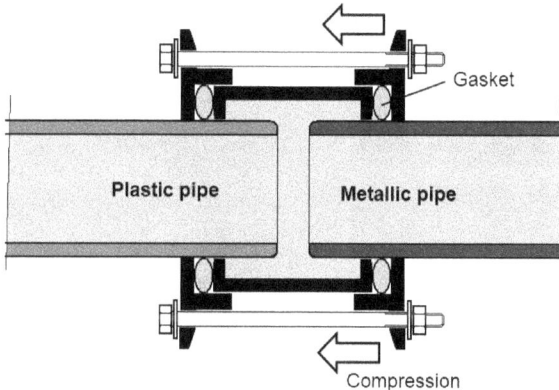

This system allows the joining of pipes of differing external diameters, generally plastic and metal, to join to an accessory (a valve in the right hand image).

They are called Gibaults or Couplings. It´s very important to specify the tolerance: that means the maximum and minimum diameter range with which an accessory can be adapted to a pipe.

Check valve

Used to limit the flow in one direction only, via a gate or spring mechanism. They can be clearly identified with an arrow marking, denoting the permitted direction of flow. The only other accessory which has the same marking is a meter. In the image, the non-return valve at the mouth of a borehole prevents backwards flow once the pump has stopped.

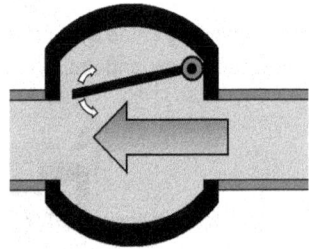

Gate valve

Consists of a disc perpendicular to the flow, which is raised or lowered via a wheel, joined to a screw. They are used to shut off the flow and are usually fully open or fully shut. They cannot be shut off suddenly, which helps to avoid the water hammer effect. It´s important to design a system so that certain points can be isolated in case of a breakage, thereby limiting the amount of water lost. A 1km section of 200mm pipe alone holds 125,000 litres.

Depending on the area a breakage can result in flooding and make repair impossible. Placing valves every 500m helps to locate blockages and saves a lot of time. Trying to find a blockage in a 6km section can be exasperating.

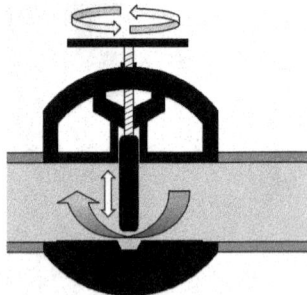

Ball valve

This consists of a ball situated in the line of flow, with a hole drilled through the middle in one direction. When this *tunnel* is swivelled in the direction of flow, water is allowed to pass. When it´s turned perpendicular, the flow is shut off completely. This kind of valve can be shut off very quickly with a ¼ turn, causing a water hammer. For this reason, they are common for smaller pipe diameters, and are cheaper and more robust than their gate valve equivalents.

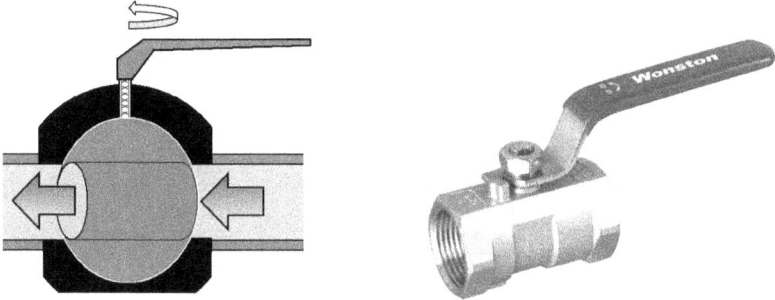

Float valve

Consists of a float joined to a rod. When a container is empty, the float drops and it´s weight opens the valve. As the water level rises, the float gradually closes the valve. They are used to avoid over filling tanks, placed at the tank inlet, at the end of a pipeline. You can find them in toilet cisterns. Above certain diameters, it can be very difficult to find these kinds of valves.

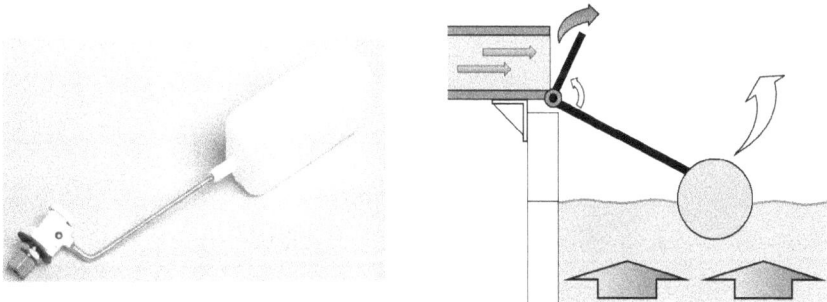

Air valve

The simplest models consist of a chamber with a float. When there is no air in the pipe, the float seals the exit hole. When there is air, the float drops, opening the hole and allowing it to escape. This kind of model shuts off suddenly causing a water hammer. When you order them, try and get hold of the double chamber models (antishock double chamber). They are expensive, and sometimes difficult to get hold of. In general, try to avoid them, as described in Chapter 4.

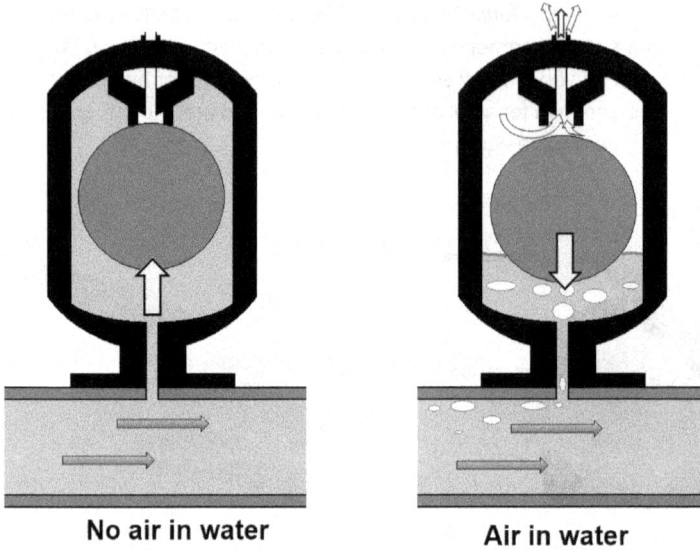

No air in water   **Air in water**

Saddle bust

These are accessories which allow a smaller pipe to be connected to a mainline for a domestic connection or public tap stand. They serve the same function as a T, and consist of a brace with a gasket which is bolted onto the main pipe. Then a hole is drilled in the centre of the threaded protruding area to allow the water through, to where the domestic connection pipe is joined. **In projects with a lot of connections, it´s vital that they are of good quality, to avoid multiple leaks which are tricky to fix.**

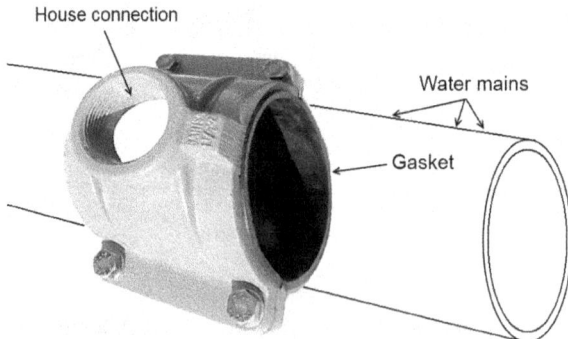

House connection

Water mains

Gasket

The connections tend to be organised around a basic *connection kit* which consist of all the necessary accessories. An example can be seen below. Note that the connection itself is made on the upper part of the pipe, to avoid the passing of sediment and to allow for air to escape:

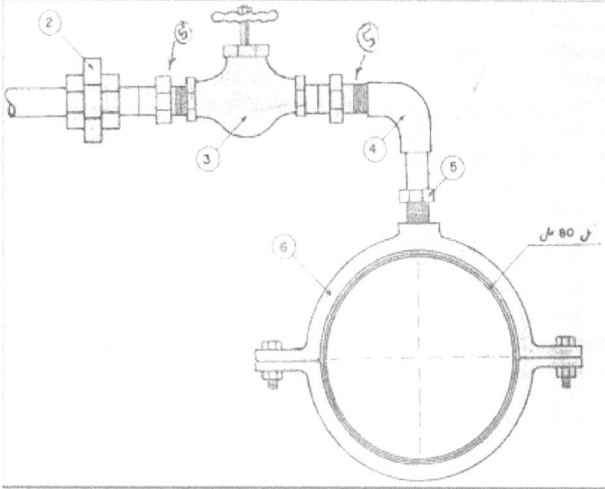

Non-essential accessories

Flow-reducing valves, pressure-maintaining valves, pressure-reducing valves and a variety of other complicated valves are delicate, expensive, and difficult to get hold of in many areas. These kinds of valves are easily recognised by their complexity, and often look a little "Sputnik" as an Afghan operator once joked: 16 years later he still didn´t know what function it served or whether or not it worked:

Why install a pressure-reducing valve at a cost of € 3,700 for 200mm if a concrete break pressure tank costs 7 times less and is much more robust?

Logistics and accessory quality

To order an accessory, 4 parameters must be mentioned: type of accessory, diameter, nominal pressure and type of joint. For pipe accessories, the type of material too, for example:

T of PVC, 100/50mm, class D, glued.
Gate valve, 100mm, 10 bar, flange according to PN standard.

- The variety of accessories can be enormous. Don´t insist on getting exactly the one you´ve ordered, but rather, one that serves the same function.

- Include all the possible drawings and diagrams so that the person who is purchasing can make themselves understood to the supplier. Since human nature is an old friend, unintelligible orders tend to pile up in desks indefinitely. On the other hand, include second and third options in case the material is not available:

ANEXO 1. Detalles gibault

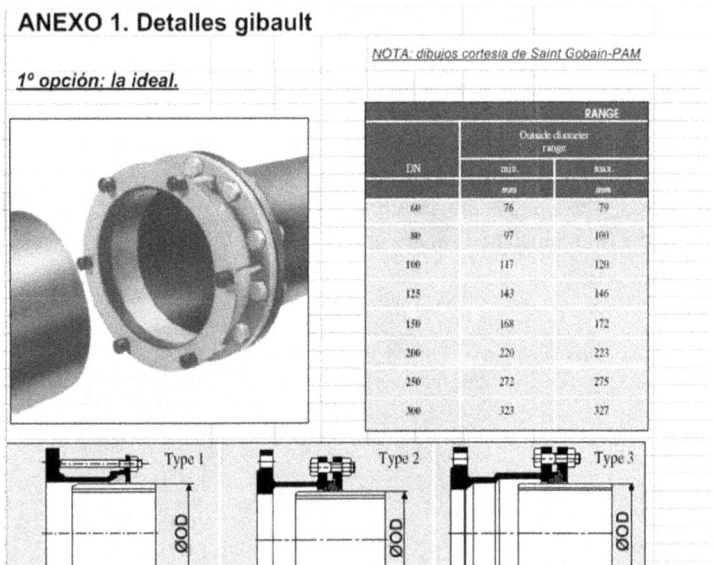

NOTA: dibujos cortesia de Saint Gobain-PAM

1º opción: la ideal.

	RANGE	
	Outside diameter range	
DN	min. mm	max. mm
60	76	79
80	97	100
100	117	120
125	143	146
150	168	172
200	220	223
250	272	275
300	323	327

Type 1   Type 2   Type 3

Fig 9.3. Complementary explanations for the purchase of a Gibault in Peru.

- Check availability on the local market to avoid having to place orders from Europe, which can take months.

Regarding quality:

- Avoid cheap Chinese accessories. Even if they are marked with "Italy" or "England" they can be easily distinguished due to their poor finishes. Note the irregular outline, lumps, bubbles and burrs on this gate valve:

- At all cost avoid low quality tools. They not only end up being more expensive, but they can injure the workers. Note the air bubbles, cracks and texture of this *"steel"*:

10. Cement construction

10. 1 INTRODUCTION

The ins and outs of construction with cement exceed the objectives of this book. In this chapter the different types, requirements and most common mistakes are included, to provide an understanding of the processes, for supervision of the basic aspects and for organising materials and labour.

Construction projects are dangerous. If you don´t have specific training, leave it in someone else´s hands.

10. 2 CEMENT

The cement normally used in construction is Portland cement. Price varies greatly from one country to the other, but it´s usually between 3 and 9 Euros per bag or sack of 50kg. The quantity of cement a country produces is one of the development indices.

Aggregates

Cement on its own has few applications and would be expensive. It´s normally used with aggregates: sand, gravel or stone. Aggregates occupy a large percentage of the volume, allowing for a considerable saving of cement. Cement acts as the matrix which holds together and provides cohesion to the aggregates. Depending on which type of aggregate is used and in what proportions, concrete, mortars and masonry are produced.

Hardening

Mixing water with cement in the correct proportions produces a paste which can be moulded and hardens over time. It can be worked and shaped comfortably for between 30 to 60 minutes. After 4 hours it can no longer be worked.

Storage

As the reaction which causes cement to harden is a hydration reaction, cement must be protected from moisture for it to maintain its properties, stored in a dry place, raised above ground and covered in plastic. For the same reason, other than for small jobs, it should be mixed by bag. Stacking cement more than 4 bags high is favourable for hardening, and stacking bags in a compact formation limits the circulation of humidity. In normal conditions cement is a grey powder with a consistency similar to flour. When cement has deteriorated it forms lumps and granules. Cement deteriorates over time in storage. It should be stored in such a way that the first to come in is the first to leave. After 3 months of storage, it will have lost around 20% of its strength; after 2 years, 50%. If you´re using cement more than 6 months old, use it for less delicate applications or increase the quantity of cement in the mix by between 50% and 100%.

Logistics

Each sack of cement requires between 28 and 33 litres of water (depending on the water content of the sand aggregate). Unless you have a supply of water near the construction site, you will need to organise and budget for water supply.

10. 3 LIME

For some applications which are not too demanding, lime can be substituted for cement. One interesting application is for water proofing small leaks in tanks. A whitewash of lime, 5kg for every m^3 of water, will seal leaks. Lime hydration generates considerable heat, so precautions must be taken to avoid burns.

10. 4 MORTAR

This is a mix of cement and sand. It´s used for masonry or brickwork, and in thin layers for plastering. Depending on the proportion, always by weight, it has distinct uses:

- **1:4** (parts cement : parts sand). This is the general use mix, for brickwork. Brickwork is a way of building by interspersing blocks, for example, in a brick wall.
- **1:3.** Rough plastering for walls to protect them from humidity.
- **1:2.** Final plastering for walls.
- **1:1-1.5.** Used for inserting anchors in rock or around GI pipes going through walls.

The total volume of mortar produced is that of the sand which has been used. Depending on the type of mortar, the quantities of each material and the labour required are approximately:

1 m³	1:4	1:3	1:2	1:1	
Cement	7.2	9.5	14.4	28.8	50 kg bag
Sand		1			m³
Unskilled labour		4			Man-days
Skilled labour		1.1			Man-days

10. 5 STONE MASONRY

These are walls built with a 1:4 mortar and stone. The resulting walls are thick, as it´s not practical to build walls less than 50cm thick.

Dry set masonry

Medium sized stones are joined with mortar. Spaces are left between stones to save on material and providing a high degree of permeability. It is used to protect exposed material or foundations, among other things. The volume is that of the stone used, and requires 7.2 bags of cement per m³ for a proportion of 80% stone : 20% mortar.

Irregular stone masonry

Similar to the one above, but special care must be taken to fill all the gaps between the stones. To do this, the proportion of mortar is increased by up to 35%. The stones are laid without cutting such that they fit together leaving the smallest possible gaps between each. However, no stone should reach from one side of the wall to the other.

The quantity of material and labour required is approximately:

$1 m^3$		
Cement	2.5	50 kg bag
Sand	0.35	m³
Stone	0.65	m³
Unskilled labour	3.2	Man-days
Skilled labour	1.4	Man-days

Regular stone masonry

Stones are cut and laid in order, like in a castle, requiring 70% cut stone and 30% mortar.

The quantity of each material and required labour is approximately:

$1 m^3$		
Cement	2.16	50 kg bag
Sand	0.3	m³
Stone	0.7	m³
Unskilled labour	5	Man-days
Skilled labour	2.8	Man-days

10.6 BRICKWORK

This is the typical brick wall. Locally made bricks vary in dimensions. They are usually around 20x10x5cm, although any size is possible. The construction technique for these kinds of walls is well-known locally. Nonetheless, pay special attention to these points:

• The quality of bricks can vary and insufficient firing can mean they break easily.

• In long walls, supports are required for stability.

- Brick walls tend to rot when they are in direct contact with the ground. To avoid humidity, they are often built on a base of stone masonry. To finish they are plastered with mortar.

Brickwork is 75% brick and 25% mortar. The quantity of each material and required labour is approximately:

1m³		
Cement	1.8	50 kg bag
Sand	0.25	m³
Brick	0.75	m³
Unskilled labour	2.8	Man-days
Skilled labour	1.4	Man-days

10. 7 CONCRETE

Concrete is a mixture of cement, sand (>5mm) and gravel (>20mm) in varying proportions, so a concrete mix of 1:2:4 has one part cement, two sand and four gravel. The units are always expressed by weight, since mixing by volume is incorrect. The more cement, the greater resistance. The most common mix is 1:2:4.

Hardening

Concrete has sufficient resistance to be put into use after 7 days, in the majority of applications. Definitive hardening is reached after 28 days, following which it continues hardening very slowly.

The amount of water is key in the final resistance of concrete and cement in general. To avoid evaporation, it´s important to cover the structures and wet them periodically during the first week of hardening.

The speed of hardening reduces with temperature and is halted with freezing. In cold climates, structures should be protected from the wind and cold, using warm water for mixing and increasing hardening times. Working below 0°C requires using additives.

Formwork and reinforcing

Recently mixed concrete has the consistency of a paste. To hold it in place and give it the required shape a mould or form is used, made of wood or metal.

Fig 10.6a. Wooden formwork for a break pressure tank.

Concrete is highly resistant in compression but weak in tension. To improve its strength steel bars are incorporated, which absorb the tensile forces. The end result is reinforced concrete.

Fig 10.6b. Verification of the reinforcing for a water tank.

Material and labour estimates

With concrete there are 3 fundamental activities: formwork, reinforcing, and mixing.

The required quantities for the <u>mix</u> and the labour are approximately:

1m³	1:1:2	1:2:4	1:3:6	1:4:8	
Cement	10.72	6.52	4.56	3.5	50 kg bag
Sand	0.37	0.45	0.47	0.48	m³
Gravel	0.74	0.9	0.94	0.96	m³
Unskilled labour		4			Man-days
Skilled labour		1.1			Man-days

The quantity of <u>steel</u> or <u>rebar</u> will be determined by the design. Steel is generally used and quoted in kg. The relation between kg and linear meters for each diameter is:

Bar diam. mm	3.4	4.2	5	6	7	8	9	10	11	12	16	20	25	32
Kg/m	0.07	0.11	0.15	0.22	0.3	0.4	0.5	0.62	0.75	0.89	1.58	2.47	3.85	6.31

To bend steel into the required shape and build the formwork (which varies according to the complexity of the structure) requires approximately 0.004 working days of a rebar worker and 0.008 working days of non-specialised labour for every kg of steel.

When it comes to preparing a budget, it´s easier to calculate the number of kilos of steel per m³ of concrete than working out each individual bar in a given design.

The <u>formwork</u> depends on the shape of the structure and the local availability of materials. This tends to be budgeted with a global sum.

Common mistakes

- **Drying out of the structures.** Those that are left exposed to the sun and wind will lose moisture very quickly. The hardening process lacks sufficient water and the resulting concrete is brittle. Make sure the structures are covered and periodically watered.

- **Pouring concrete**. When concrete falls and hits the ground, the gravel advances through the liquid mix and the components become separated. The cement is left concentrated in the upper area, followed by the sand, and the gravel piles up below.

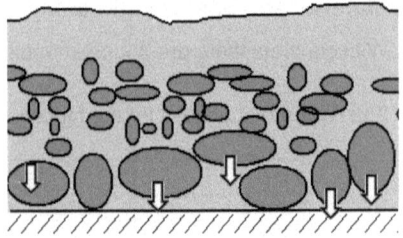

- **Mixing on the ground**. Mixing concrete directly on the ground causes it to lose water and gather impurities. Mixing should be done on metal sheets or concrete slabs.

- **Not vibrating**. So that the concrete reaches all areas and to avoid air bubbles, it should be vibrated with a machine or stirred inside with a bar, hitting the formwork.

- **Superficial reinforcing**. So that the reinforcing is sufficiently integrated into the concrete, it should be covered by at least 2.5cm. Under certain pressures, the rebar in the photo (7mm) will be pulled out of the concrete. To guarantee adequate distances, spacers are used.

As the rebar is what gives concrete its strength and prevents sudden collapse, inspection is particularly important. They cannot be welded, unless they are done so industrially. Establish in the contracts that before pouring concrete the contractor should inspect the reinforcing. Trying to save money on steel or using inadequate steel is potentially very dangerous. The presence of small quantities of rust on the rebar is normal, and can even improve adherence between the cement and the steel.

10. 8   PERMEABILITY AND WATER PROOFING

Water proofing of concrete is achieved with additives or a layer of bitumen.

Bitumen layer

An alternative method is to plaster with 2 layers of mortar, 1cm thick, with a progressively greater proportion of cement: 1:4, 1:3, 1:3, 1:1. The smooth finish will greatly assist in water tightness. A smooth finish or using metal formwork reduces the number of pores and leaks.

For quick sealing of leaks, a lime whitewash can be used of $5kg/m^3$ (see section 10.3).

Permeable concrete

This is a kind of concrete which lets up to $90l/m^2*s$ pass through a thickness of 10cm. It's very useful in avoiding puddles and pools around public tap stands, and for intake works.

Basically, it's concrete without sand, using slightly smaller gravel (6-10mm). The mix by volume (cement : fine gravel : water) is 1: 4.5 : 0.6. It can be reinforced, and has strength similar to traditional concrete if it's made correctly. The quantity of water is critical: too much causes cement membranes which make it water proof...too little and it becomes brittle.

Fig 10.7. Detail of a permeable concrete ring. The gravel is the size of the pencil.

11. Intake works

11. 1 INTRODUCTION

The intake works are the part of the system which collects the water and introduces it into the pipe. Each situation requires a specific focus and there is no infallible design which can be repeated universally. Here there is a selection to give you ideas to find a solution that is best adapted to your situation.

Well-designed intake works should perform certain functions besides the obvious one of collecting water:

- Condition the water, reducing the quantity of suspended material. In fact, a well thought out intake can completely eliminate the need for subsequent treatment.
- Protect the water from contamination, preventing the entrance of animals, manipulation of any kind and risks from flooding.

11. 2 CHOOSING A LOCATION

The only limitation is that it be in direct contact with the source, which leaves a wide range of possibilities.

Certain criteria

For springs the intake is built where the spring rises. In the case of streams, there are more possibilities, and it´s important to take into account that:

1. The intake should be accessible throughout the year.

2. The choice of location should protect the intake from damage due to flooding and mud slides.

3. The potential for erosion should be small. Rocky areas or where the water flows at low velocities are more likely to be affected by erosion.

4. The location should be upstream of sources of contamination: washing areas, settlements, water holes for animals...

5. The pipe should be free to exit the intake without following the stream, for several meters until it can be sufficiently pressurised. The top of a slope is a good place for this.

6. Take advantage of cavities, pools and narrow stretches.

7. Laying foundations is simple, so that lateral and lower level leaks are avoided. As far as possible, there should be contact with the bed rock.

Pre-intakes

It´s not necessary to have the entire intake in the same place. If there are areas further from the source that are more favourable, with space to build filters or sedimentation tanks, or better protected, a pre-intake can be built in the form of a protected pipe intake:

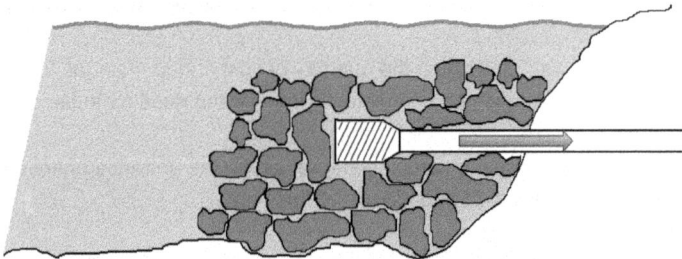

By adding a valve box for control, this is the simplest form of intake. The stones around a slotted end-pipe provide physical protection to the pipe and act as a roughing filter. The pipe should be at least 30cm below water at all times, to avoid the entrance of floating particles such as organic material and excrement, which can enter the pipe and block it. Moreover, the water near the surface has algae and other organisms which tend to give off a bad smell and taste. On the other hand, the intake should not be too close to the bottom to avoid the entrance of sedimented contaminants.

11. 3 STREAM INTAKES

The challenge is to get sufficient depth so that the water flows without any problems, which often requires the water level being raised with small dams. Look for advantageous sites, narrow stretches, and changes in elevation. A dam creates an area of water at rest, which allows for suspended particles to sediment.

Small dams

Dams are dangerous. Unless you have specific training don´t try building dams other than small barrages a few meters long, and far from any settlements. When the dam is full it should not flood the surrounding area.

A simple way of raising the water level is to make rock barriers without cement or gabions. The rocks provide strong resistance to the flow and the water level rises to get over them:

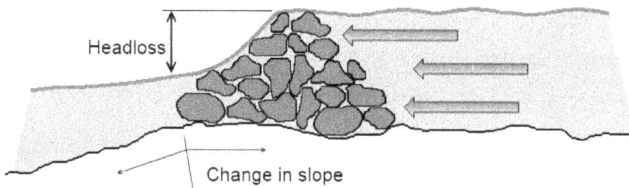

A partial dam will have the same effect. In this case, the water accelerates through the narrow stretch, and protection from erosion must be provided in the opposite bank, using rocks:

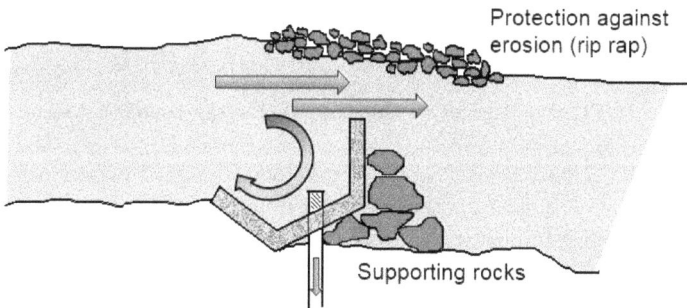

A total dam closes off the flow from one side to the other, allowing it to pass through an overflow.

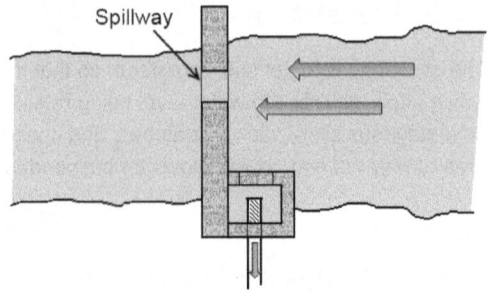

The image shows a dam for a gravity fed intake. Note how the sides are protected against erosion and enlarged by two walls made from stone masonry. The overflow is situated in the centre, and the intake, immediately to the right of the person, has a concrete slab as a lid. This is the best option, allowing for a perfect seal. The main problem with dams is leaks between the dam and the substrate. Erosion begins to open a gap until the rock disintegrates. This dam has made use of a small natural waterfall, to get down to the bedrock, and allows for rapid pressurising of the pipe given the available slope.

Fig 11.3. Dam for a gravity fed intake works, Mtabila II, Tanzania.

For the supervision of these kinds of works, dams made from stone masonry should be less than 2m high. The sides in contact with the water should be vertical and water proofed, with the opposite side sloped so that the base measures at least 0.7 times more than the height. These kinds of dams should be built on solid rock or foundations. On the side where the water runs off downstream, stones should be placed to avoid erosion which can cause the dam to collapse. Make sure the join between the rock and the masonry is tight and rough.

Infiltration galleries

This is probably the intake of choice, as the entire structure is buried, greatly reducing or even completely obviating the need for subsequent treatment. The disadvantage is that the low-lying position can make the exit and sufficient pressurising difficult, as well as it being difficult to work in flooded areas.

In its simplest form, it consists of a perforated pipe, buried in the bed of a stream or body of water. The pipe is covered in layers of gravel, progressively smaller in diameter, with sand on top. This configuration provides filtering and a considerable reduction in the quantity of bacteria in the water.

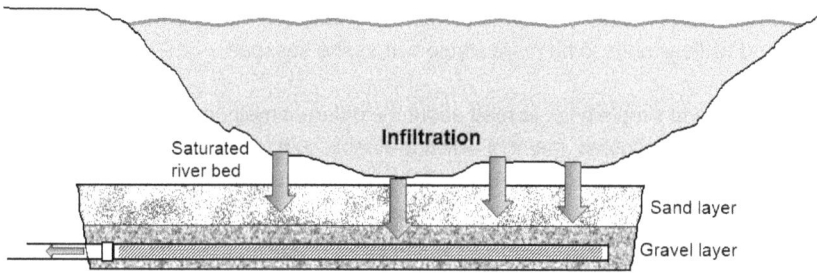

In traditional infiltration gallery designs, where the water is extracted from a well to one side, it is recommended to bury the pipe 1m into the bed. This makes the exit more complicated. In gravity flow designs, the pipe can be placed parallel with the stream flow, without the pipe exiting too low. Note that the new diagram is similar, but instead of a section view it is seen from above:

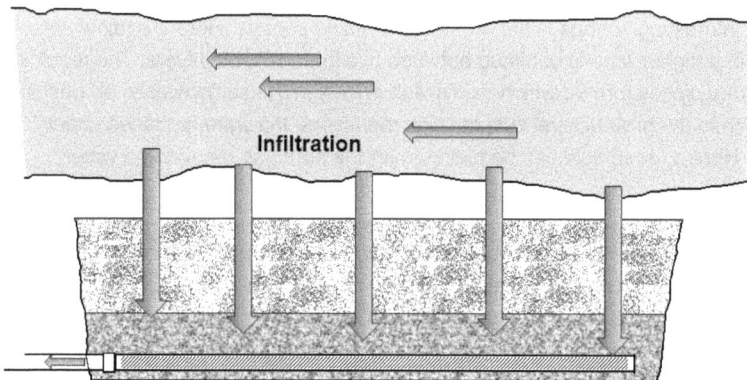

The gravel around the pipe should be similar to that used in filters for boreholes, less then 10mm in diameter.

11. 4   SPRING INTAKES

Springs are ideal sources for gravity flow water supply. If they are adequately protected the water needs no treatment, the spring tank can be extremely simple and cheap, and its flow tends to be more stable across the seasons.

An interesting and simple book to read about the different methods of tapping underground water sources, many of which are applicable to gravity systems, is *Investigación y Alumbramiento de Aguas Subterráneas* (reference 9 in the bibliography). Unfortunately it is only available in Spanish. Despite being more than 100 years old, the techniques described are excellent and second hand copies can be found on the internet or at reduced price.

Springs in flat areas

These kinds of springs are hard to use for gravity projects because they don't allow the pipe to be pressurised. Sometimes it's possible to reach a sloped area with a certain amount of digging. Another alternative, if they are at the foot of a hill, is to try and locate the underground flow by excavating. Most of these methods are described in reference 9.

The most interesting springs in flat areas are artesian springs. Here the rainwater collected in a higher area is confined between two impermeable layers. The result is that a *natural pipe* is formed which works with gravity in the same way as an artificial system. When the ground level cuts through the layers, the water surfaces under pressure. Here a receptacle can be placed over the spring to recover the water.

This is practical depending on the size of the spring. A concrete box with a lid and no floor is placed directly on top. The bottom is covered with gravel so that the water can pass through and the walls are sealed at their lowest point with a pliable mixture of clay and water. Finally, the box is backfilled with excavated material so that the water flows towards the outside, and a concrete ramp is made to evacuate the unused water, in the form of an overflow.

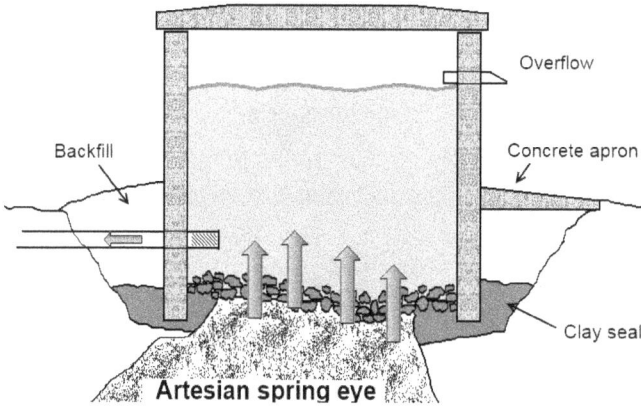

Artesian spring eye

Springs on slopes

These springs are easier to use. Basically there are two models, the simple wall (first illustration) and the spring box (second illustration). In the latter, the box is built to act as a small reservoir tank and is sized as such, although sometimes it´s not practical or possible to build it near the spring if they are large.

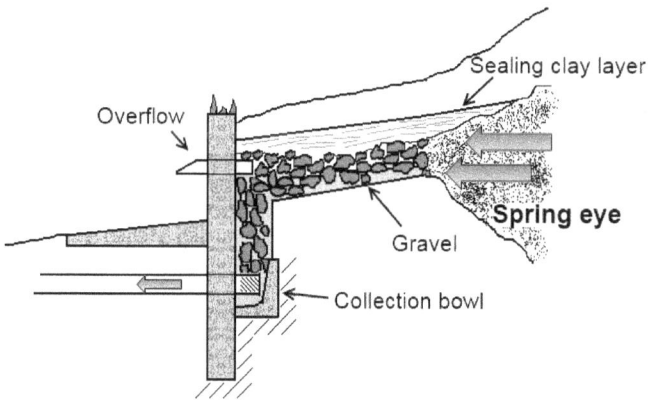

The layer of clay helps to protect the spring water from possible contamination at ground level. In the upper wall area stones are sometimes placed, on-end, pointing upwards, to stop people walking over them. The surrounding area should be fenced off to stop animals entering, and a semi-circular bank shaped according to the slope is made to stop surface water entering the spring intake. Although some manuals recommend digging a trench for this, there is a danger of reaching the water level when excavating.

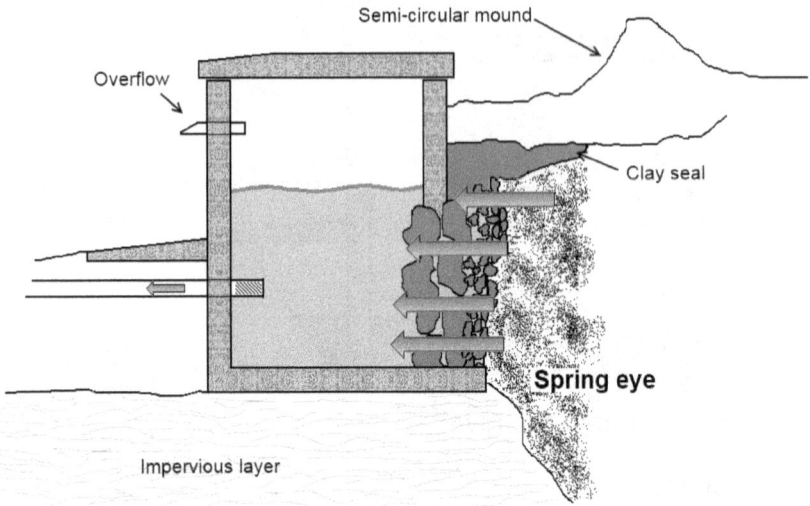

There are techniques for increasing spring flows. Generally they are based on the idea of increasing the amount of infiltration by making superficial water flow more difficult. Vegetation is planted or stone walls are raised, for example, to increase the contact time and allow more water to filter down into the subsoil.

For more information about springs, see reference 15.

Fig. 11.4 The Tempisque spring, El Salvador, with a flow of 11 l/s.

333332

11. 5   SMALL LAKE INTAKES

The main advantage here is the volume of water that's stored, which avoids the need for reservoir tanks, and even sedimentation tanks. An interesting construction in these situations is an infiltration gallery, which can be built dry next to the lake and then dug out to allow access to the water source.

In these cases, it's important to put in a scaled ruler to be able to monitor water levels.

11. 6   VALVE BOXES

Up until now these have been left out for the sake of simplicity. The ability to shut-off or regulate the flow in certain pipe sections is very important. In case of breakages, a valve can be shut, allowing the system to be drained so as to do dry repair work. Doing repairs without cutting off the flow is tricky, uncomfortable and requires bilge pumps, despite the generous smile of the worker in the photo.

Another important function of valve boxes is to release air which rises through the pipelines to the upper areas of the system. For this, a simple snorkel-like pipe is placed above the maximum possible level of the water, in the spring box:

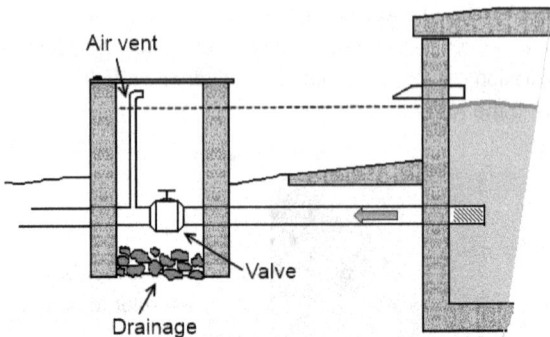

The floor of valve boxes is a layer of thick gravel which allows the evacuation of small leaks and any splashes from the air valve, avoiding flooding. If the pipe is surrounded by stagnant water, the spring becomes a source of contamination.

Valve boxes are dealt with in more depth in the next chapter.

11. 7 SEDIMENTATION TANKS

Surface water tends to contain many suspended solids which give it a turbid appearance. Aside from the appearance, these particles are problematic in terms of the smell and taste they leave in the water and the wear they cause on the pipes and accessories. Once inside the pipes, they sediment and accumulate in the low points, reducing the effective diameter.

If water is allowed to settle for some time, these particles settle and the end result is clearer water. Sand has a sedimentation velocity in water of 0.5 to 6 m/min. Finer particles and bacteria do not sediment. For practical purposes, when water is at rest for an hour, it has lost the majority of suspended solids.

In systems with reservoir tanks, a small sedimentation tank with a detention time of 15 minutes allows for the largest, more abrasive particles to sediment out before entering the pipe. The remaining sedimentation will take place in the reservoir tank. A system with no storage tanks require times closer to an hour.

Measure of turbidity

Turbidity is measured in a transparent tube of water. At the bottom there is a cross or circle 2mm wide. The water is emptied from the tube until the marking at the bottom can be seen clearly. The tube is calibrated and allows a reading to be made. The unit of measurement is NTU. For treated water, turbidity should be less than 5 NTU.

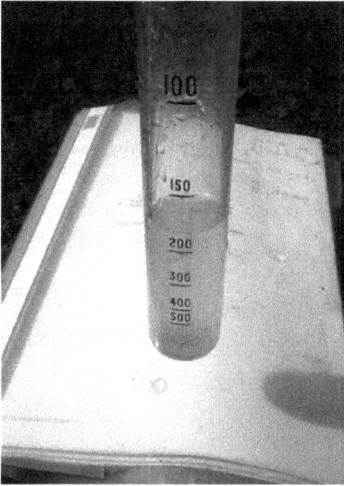

The reading in the photo is approximately 190 NTU. Note that you can see the cross on the floor of the tube, and that the measurement is taken in daylight over a white background. This sample, taken from a public tap stand of a gravity flow system in Tanzania, has an unacceptably high turbidity and indicates the absence of effective filtration.

Calculations for a sedimentation tank

The goal is to have a tank large enough to allow the water to remain there sufficient time for sedimentation to take place: 60 minutes for systems with no reservoir tank, and 15 minutes for systems with. The volume of the tank in m^3 is determined by:

$$V = 3.6 * Q * t$$ Q, flow in l/s; t, retention time in hours.

Nonetheless, not every kind of tank works for sedimentation. The internal velocity should be less than 0.005 m/s to avoid currents which impede sedimentation, and the length should be at least four times the width, to absorb incoming turbulence. To calculate the velocity:

$$v = 1,000 * Q / A * h$$ A, width in meters; h, depth in meters.

Calculation example:

Work out the dimensions of a sedimentation tank from an intake, feeding a gravity flow system with no reservoir tank, with an average future population demand that is estimated to be around 2 l/s. Work out the same for a system with a reservoir tank.

No reservoir tank:

The depth of the tank is established as 0.75m, according to the data in the section below.

The temporal demand variations will affect the input velocity at the intake. In the absence of data on temporal variations, take a value four times higher as the peak flow rate (see section 2.8):

$$Q= 2l/s*4 = 8 \text{ l/s}$$

The required volume is: $V = 3.6 * Q * t = 3.6* 8l/s * 1h = 28.8m^3$

For a velocity of 0.005 m/s or less:

$$v = Q / 100 * A * h \rightarrow A = Q /1000 * v * h$$
$$A= 8l/s / 1.000 * 0.005 \text{ m/s} * 0.75m = 2.13m \text{ or greater: } 2.2m.$$

The required tank length for 28.8m³ is:

$$L = V / A * h = 28.8m^3 / 2.2m * 0.75m = 17.45m$$

Lastly, the length, 17.45m, is verified to be more than 4 times the width, 4 * 2.2m = 8.8m.

The required tank measures 17.45m long, 2.2m wide, with a depth of 0.75m.

With reservoir tank:

With storage, the detention time is 0.25 hours. Also, as the tank absorbs temporal variations, the input flow used is the original 2 l/s.

The required volume is: $V = 3.6 * Q * t = 3.6* 2l/s * 1h = 7.2m^3$

The width is $A = Q /1000 * v * h = 2l/s / 1000 * 0.005 \text{ m/s} * 0.75m = 0.53m$

Taking the width as 0.6m, the required length for 7.2m^3 is:

$$L = V / A * h = 7.2m^3 / 0.6m * 0.75m = 16m$$

Again, it is more than four times the width.

The required tank measures 16m long, 0.6m wide, with a depth of 0.75m.

Note that the second tank is smaller and cheaper. The money saved could be used to build a reservoir tank.

A length of 16m is important and not all locations will allow for this. This can be achieved by partitioning the tank, providing considerable material savings.

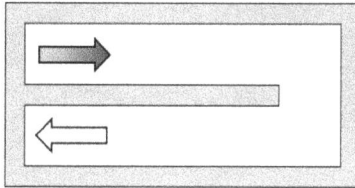

Construction details

- The turbid input water should enter at mid-height, evenly dispersed over the entire width of the tank. This can be achieved by perforating small holes in the pipe at regular intervals.

- The clean water exit pipe should be as high as possible, just below the overflow pipe.

- The optimum depth is between 0.7m and 1m.

- There should be a washout to allow for complete emptying of the tank for cleaning, together with a valve to shut off the incoming flow.

It´s fairly easy to mix the concepts and end up making a hybrid intake, which fuses a standard intake with a sedimentation tank. In the intake shown in the photo, the sedimentation tank doesn´t have the required dimensions or shape. The water comes in from above, and is taken out from below. The input flow goes directly to the intake (in the foreground) where sedimentation occurs. Finally, the construction doesn´t allow for cleaning and fails to protect the valve box from interference.

Fig. 11.6 False sedimentation tank.

12. Other components

12. 1   RESERVOIR TANKS

The tank is the most visible part of the system, and one which is a great source of pride. It´s also the ideal place to inscribe not only the name of the project and the date, but even the names of those who have taken part in the construction, or a mural depicting how to prepare a rehydration solution. The natural tendency is to want to make the tank as big as possible. In this section, you will see how to calculate the required tank size, to avoid ending up with small tanks which are always empty and oversized ones which never fill.

The principal function of the tank is to absorb the disparities between what comes in from the source and what leaves the tank, according to the varying temporal demands of the local population, so that there is always water available. If the source provides a greater supply than the peak demand, a tank is not normally required. In section 2.8 and in previous sections, there is an example of demand calculations, taking into account daily, weekly, and monthly variations, and unaccounted for water demand, based on future population.

If peak demand exceeds supply, you´ll be dealing with one of the two following cases:

a.   If demand exceeds supply now, you have limited room to manoeuvre. Simply work out the required tank size and organise the construction.

b. If demand exceeds supply from the source 20 years ahead for the projected population, the tank construction can be postponed for a future date.

When the distances are large, it can be cheaper to build a small tank and install smaller pipes between the source and the tank.

Danger

In gravity flow water projects it´s common for the reservoir tank to be built just above the houses. If the tank falls apart, the resulting tide of water will head straight for the houses. Avoid emergency tanks or corrugated iron tanks. These were developed for agricultural purposes, and not for installation near population centres. Take into account the risk from earthquakes, armed conflict and soil stability:

Fig 12.1. Tank installations right on top of earthquake fissures.

Building a tank in a dangerous place requires attention and supervision from qualified personnel. This is not something to build *with the community* or from plans with unspecialised labour.

Working out the required size

The simplest way of working out the tank size is to take a balance of the inflow and outflow, similar to a kind of accounting exercise. Over a period of 24 hours, the required tank size is the maximum registered volume minus the minimum registered volume. If there are various demand scenarios, these must be taken into account. This one, for example, is a calculation which takes into account the demand from both animals and people. The maximum volume (in the TOTAL column) is 3500 litres, and the minimum is -2688.

	People	Animals	Total demand	Supply	Balance	TOTAL
0:00	0	0	0	700	700	700
1:00	0	0	0	700	700	1400
2:00	0	0	0	700	700	2100
3:00	0	0	0	700	700	2800
4:00	0	0	0	700	700	**3500**
5:00	59	2184	2243	700	-1543	1957
6:00	235	2184	2419	700	-1719	238
7:00	764	2184	2948	700	-2248	-2010
8:00	1058	0	1058	700	-358	-2369
9:00	1000	0	1000	700	-300	**-2668**
10:00	412	0	412	700	288	-2380
11:00	118	0	118	700	582	-1798
12:00	59	0	59	700	641	-1156
13:00	59	0	59	700	641	-515
14:00	59	0	59	700	641	126
15:00	235	0	235	700	465	591
16:00	588	0	588	700	112	703
17:00	706	0	706	700	-6	697
18:00	353	0	353	700	347	1044
19:00	118	2184	2302	700	-1602	-557
20:00	59	2184	2243	700	-1543	-2100
21:00	0	0	0	700	700	-1400
22:00	0	0	0	700	700	-700
23:00	0	0	0	700	700	0

$$V = V_{maximum} - V_{minimum} = 3500 \text{ litres} - (-2668) \text{ litres} = 6168 \text{ litres}$$

By making a graph of the values in the TOTAL column, the water level in an imaginary tank can be visualised.

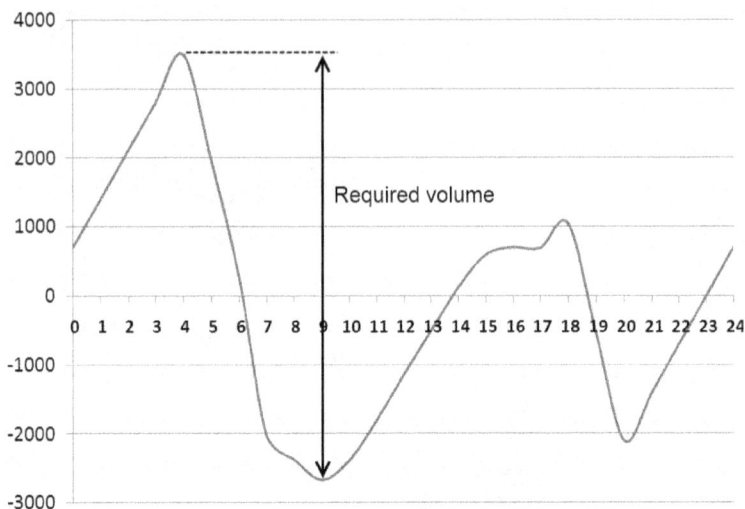

To work out the outputs for each hour, you need to have established a daily demand pattern. Section 2.5 shows how difficult it is to measure one in the field. Talking with the local population and getting to know their customs will help you put together a tentative pattern.

Note that for this method to work, the water supplied and consumed must be exactly the same, otherwise you´ll be storing water which is never consumed:

- If the source provides a constant supply, the hourly supply corresponds to the total demand over 24 hours.
- If the source is intermittent, as with pumping to the tank, it´s very important that the hours of supply (pumping) coincide with the peak demand. As most of the volume will be consumed at the same time, the required storage is less and tank size optimal.

Calculation example:

In a mixed system supplying 10,000 people with 50 litres per day, pumping is planned with a pump that delivers 50 m^3/h from a nearby river to the tank. From the tank onwards the system is gravity flow. The demand pattern is unknown. ¿What should the tank volume be?

The total daily demand is:

10,000 people * 50 litres/person*day = 500,000 litres per day or 500m^3

The required number of pumping hours is: 500m^3 / 50m^3/h = 10 hours

To establish a tentative demand schedule we can follow the guidelines in section 2.5. We can establish that there will be a demand peak of 3 hours during the morning (roughly between 6:00 and 9:00) where 50% of the water is consumed, followed by a second peak in the afternoon (15:00 to 17:00), where 20% is consumed. Pumping hours, each of 50m^3 are timed to match the two peaks:

	%	Demand	Supply	Balance	TOTAL m^3
0:00	0	0	0	0	0
1:00	0	0	0	0	0
2:00	0	0	0	0	0
3:00	0	0	0	0	0
4:00	2	10	0	-10	-10
5:00	5	25	50	25	15
6:00	13	65	50	-15	0
7:00	18	90	50	-40	-40
8:00	15	75	50	-25	-65
9:00	5	25	50	25	-40
10:00	3	15	50	35	-5
11:00	2	10	0	-10	-15
12:00	2	10	0	-10	-25
13:00	1	5	0	-5	-30
14:00	4	20	50	30	0
15:00	10	50	50	0	0
16:00	8	40	50	10	10
17:00	4	20	50	30	40
18:00	2	10	0	-10	30
19:00	2	10	0	-10	20
20:00	1	5	0	-5	15
21:00	1	5	0	-5	10
22:00	1	5	0	-5	5
23:00	1	5	0	-5	5
	100	500	500	0	700

The required volume is : V = V$_{maximum}$ − V$_{minimum}$ = 40m^3 − (-65m^3) = 105 m^3

To this should be added the appropriate reserve for fires.

Fire reserve

Section 2.7 covered the need to establish a fire reserve. The volume of this reserve can be obtained by talking to the local community and the people who would be putting out a fire. The tank construction should take this reserve into account as far as possible, so that it´s always available. The trickiest part is maintaining it during tank cleaning times. The best is to have tanks with two chambers: one chamber maintains service while the other one is cleaned.

The pipe arrangement is fundamental, so that the fire reserve cannot be consumed under normal conditions, via the main service pipe for the system.

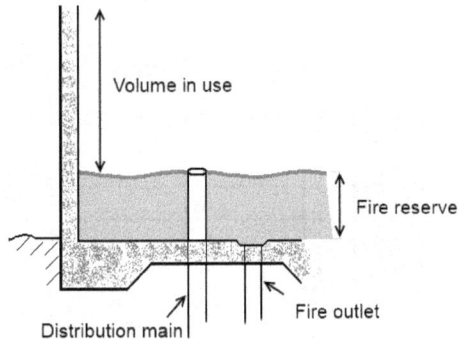

Volume in use

Fire reserve

Fire outlet

Distribution main

12. 2   HYRDRAULIC RAM PUMP

The hydraulic ram is a pump which uses the energy of falling water to pump it upwards. In gravity flow projects this opens up an interesting possibility, of delivering water to users who are above the source without the costs of electrical pumping.

Operation

The pump has a feeder drive for incoming water and a propulsion pipe for outgoing pumped water. Water which falls from a certain height gains speed until it is able to overcome the resistance of the main non-return valve. When this happens, the column of water stops abruptly and tries to find the only means of expansion possible in the air chamber. This opens the second non-return valve and from there the water passes into the propulsion pipe, pushing the entire column of water until its energy is expended.

At this moment the falling column of pumped water closes the second non-return valve, keeping it at the same height. Then the spring opens the main non-return valve and the water in the pipe passes once again through the exit, gaining momentum. Once it has reached sufficient speed to close the main non-return valve, the cycle is repeated. This is a pump which pumps by pulses. The air chamber acts as a shock absorber to reduce the pressure which the different elements are subjected to.

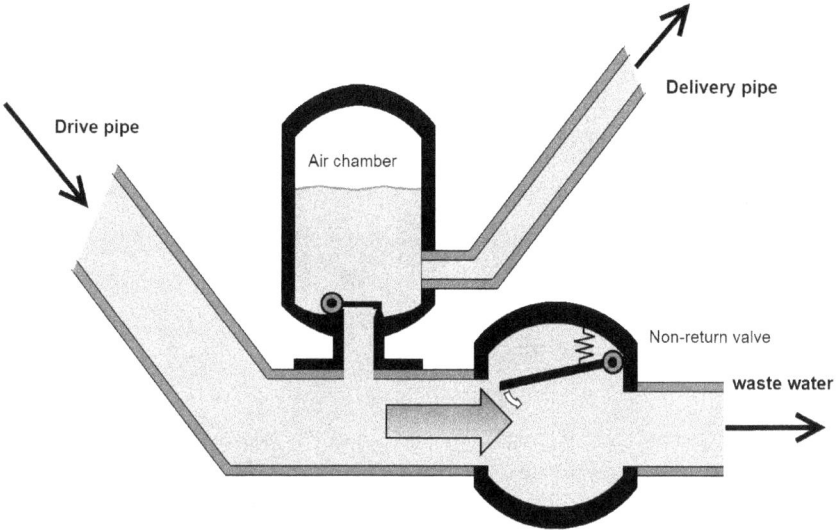

Although it´s not a good idea to use the entire flow supplied by the source, it is possible to use all the flow for a few hundred meters to power the ram pump and then return it to its original course via the washout.

Sizing

Homemade ram pumps have much lower performance than manufactured ones. To select the required pump, the best is to contact the manufacturer. You need to specify:

- How many litres per day you need to pump.
- The available input flow.
- The height to which you want to pump.
- The available height of the delivery pipe.

12. 3 BREAK PRESSURE TANKS

This is a system component designed to dissipate pressure in the system, as talked about briefly in section 5.3. Where the differences in elevations are very pronounced, the cumulative pressure puts the pipes in danger and makes the system unwieldy and dangerous.

In the same way a tyre loses pressure when it´s punctured (returning to atmospheric pressure), a pipe which empties into an open chamber also loses pressure. A break pressure tank consists of a tank with an entrance and exit pipe. The entrance pipe has a float valve to cut off the flow when there is no demand. The exit pipe is free.

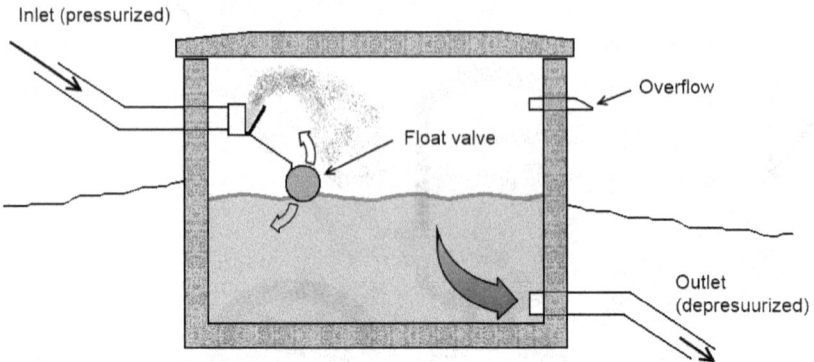

Inlet (pressurized)

Overflow

Float valve

Outlet
(depresuurized)

There is no minimum size, although a volume of 1 or 2m^3 allows for less abrupt flow regulation and increased life expectancy for the float valve. In general it´s difficult to find float valves larger than 50mm. The installation of a reduction fitting allows for the use of smaller valves.

The entrance flow can deteriorate the concrete or stone finish, gradually washing away the cement. When the tank is half full, the discharge should be falling onto water. When it´s almost empty, it should fall onto a flat stone (i.e. slate) set into the tank structure.

12. 4 AIR VALVES

Air valves have been dealt with in section 9.3 and throughout the book. They are used to expel air which accumulates in the high points of a system. However, they are expensive and difficult to get hold of, and not all the high points require one:

- Connection points resolve the problem acting as air valves.

- At the high points which coincide with points of demand, air is evacuated through the taps.

- Pipe configuration can act to eliminate air accumulation by itself.

High points closest to the hydraulic grade line are the most problematic, as the pressure is lower and air bubbles are larger at lower pressures.

Gravity Flow Water Supply   155

Evaluating the need for air valves

If a new pipe is being filled, the water passes over the first high point and falls until it reaches the first valley. When the low point has filled the entire pipe the air remains trapped in the form of an air block. The pressure which the air block is under is the difference in height between the source and the start of the blockage, and is referred to as **compressive pressure**.

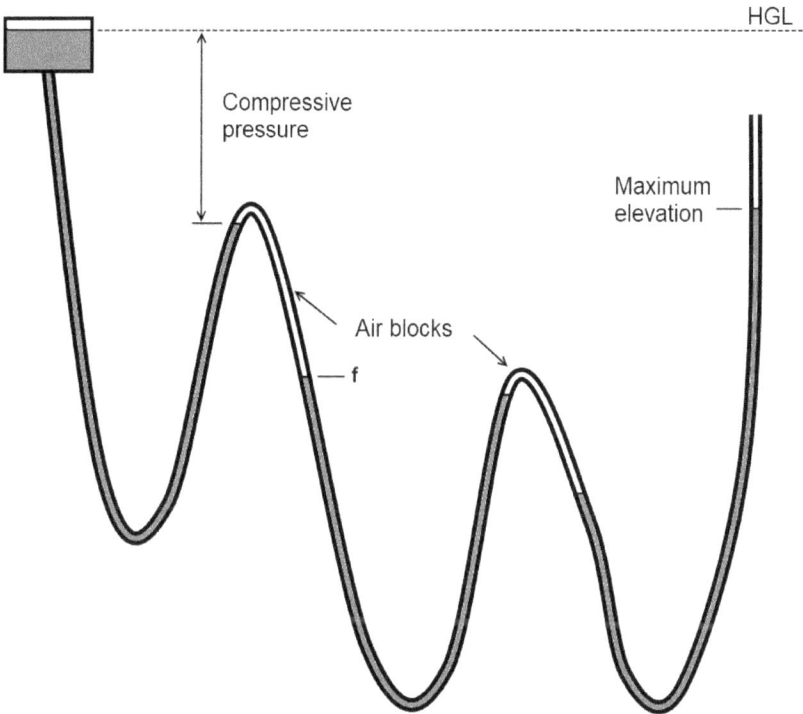

If the air block is not complete and a small flow can pass through, it will gradually be eliminated. For complete flow, the **highest elevation** must be higher than the outlet elevation. The air in the blockages is light so the elevation of air blocks is lost pressure which reduces the pressure at the maximum elevation. The process for taking into account this loss of pressure is shown below. Pay close attention, as it´s simple but can get muddled:

1. Begin by working out where the air blocks will be along the pipeline. Start with the first air block.

2. Calculate the volume of trapped air, V_0, multiplying the length between the high point A and low point B by the internal diameter (see Appendix B).

3. Work out the corresponding head loss for a minimum flow of 0.1 l/s.

© Santiago Arnalich   Arnalich. Water and habitat   www.arnalich.com

4. The head loss subtracted from the static pressure is the compressive pressure, Pc.

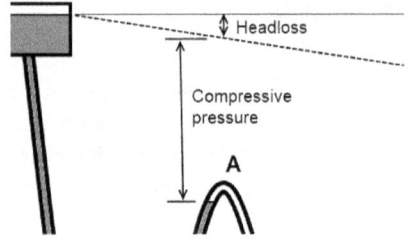
Headloss
Compressive pressure
A

5. Calculate the pressure, in kg/cm^2, in the air block: $P_1 = (0.1 * Pc) + 1$

6. Calculate the volume of the blockage: $V = V_0 / P_1$

7. If Z is the volume of a meter of pipe, calculate the length of pipe for the blockage: $L = V / Z$

8. With this length, you can locate the end of the blockage, f in the initial diagram, and work out its elevation.

9. The compressive pressure of the second blockage is the difference in height between the end of the first blockage and the start of the second, less the head loss for this section.

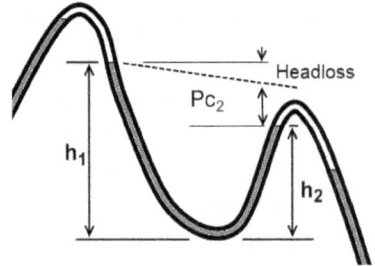
Headloss
Pc_2
h_1
h_2

10. The pressure of the second blockage, P_2, is: $P_2 = P_1 + 0,1 (h_1 - h_2)$

11. Repeat steps 6 and 7 and continue through all the blockages until you have worked out the elevation and pressure of the last air block, P_x.

12. Calculate the equivalent height of the last air block: $h_e = 10 (P_x - 1)$

13. Work out the head loss from the first air block to the end of the pipeline. Subtracting this from h_e and adding the elevation at the end of the first blockage, you obtain the maximum elevation, E_m.

If E_m is higher than the end of the pipe, the air blocks will disappear without the need for air valves. If this is not the case, try increasing the pipe diameter for the section where there are air blocks and repeat the calculations. If that still doesn't get the desired results, install an air valve at the highest air block and repeat the calculation.

Calculation example:

The calculations for the gravity flow system in the graph have shown that the required pipe for the whole section is HDPE PN10 of 63mm. Work out if air valves should be installed:

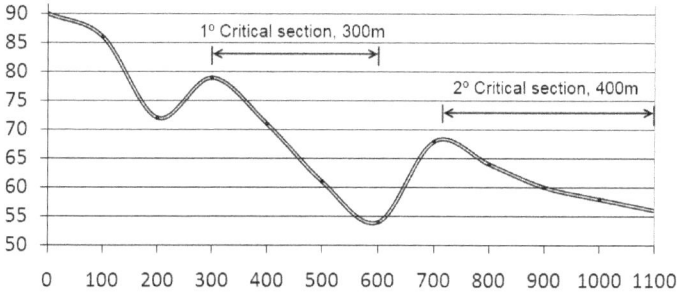

Elevation	90	86	72	79	71	61	54	68	64	60	58	56
Distance	0	100	100	100	100	100	100	100	100	100	100	100
Chainage	0	100	200	300	400	500	600	700	800	900	1000	1100

(Step 1). The critical sections are marked on the graph.

(Step 2). The internal diameter of the HDPE pipe is 55.4mm. Using decimeters to obtain litres ($1l = 1 dm^3$), the volume of air in the air block in each critical section is:

$$V \text{ per linear meter} = \pi\, d^2/4 * 1m = 3.14 * 0.544^2 dm^2/4 * 10dm = 2.4 \text{ litres}$$

$$V \text{ of the } 1^{st} \text{ critical section} = 300m * 2.4l/m = 722.8 \text{ litres}$$

$$V \text{ of the } 2^{nd} \text{ critical section} = 400m * 2.4l/m = 960 \text{ litres}$$

(Step 3). The head loss for 0.1 l/s in the pipe is 0.

(Step 4). The compressive pressure is: $Pc_1 = (90m - 78m) - 0m = 12m$.

Analysis of the first air block:

(Step 5). The pressure of the air block is: $P_1 = (0.1 * 12) + 1 = 2.2 \text{ kg/cm}^2$

(Step 6). The compressed volume of the blockage is: $V = 722.8 / 2.2 = 328.54$ litres.

(Step 7). The equivalent length is 328.54 litres / 2.4 l/m = 136.9m

(Step 8). From the profile, the elevation at the 136.9m point from the peak is around 67 m.

Analysis of the second air block (Step 11):

Pc_2 = (67m - 68m) - 0m = -1m

P_2 = (0.1 * (-1)) +1 = 0.9 kg/cm^2

V_2 = 960 / 0.9 = 1066.66 litres

L_2 = 1066.66 / 2.4 = 444.44m

The elevation of the 444.44m point downstream from the second peak is around 55m (outside the graph).

(Step 12). The equivalent height is: h_e = 10 (0.9 -1) = -1m

(Step 13). The total head loss is 0m. The maximum elevation is: E_m = -1m -0m + 55m = 54m.

As 54m < 56m, air is a problem. If a change of pipe diameter in the critical sections does not change this, an air valve should be installed at the first peak, and calculations made again. Here the example is concluded for the sake of brevity.

12. 5 WASHOUTS

Washouts are placed in the low points of a pipeline and allow it to be emptied in case of breakage or to eliminate sediment which has accumulated in the low points. The quantity of water held on a pipeline is important. A 1 km pipe of 63mm holds 24,000 litres.

On emptying, the water should be able to flow off without causing damage, either along an existing channel or into a soakaway.

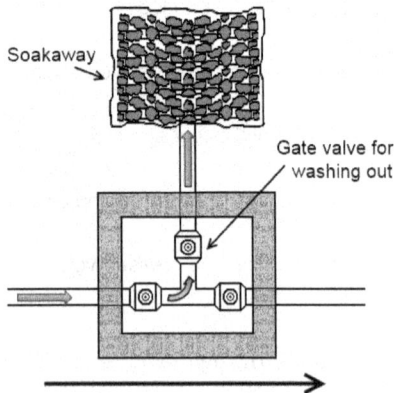

Soakaway

Gate valve for washing out

The 3 valves allow different sections of pipe to be emptied, either the right or left hand ones, or both. The flow can also be shut off in case the washout valve breaks.

12. 6 SOAKAWAYS

Stagnant water worsens the health conditions in a given area. To avoid this, waste water from washouts and public tap stands should be channelled into pits filled with rocks. There, the water filters slowly into the ground without causing health problems.

The image shows a soakaway which is connected to the overflow of a reservoir tank under construction. On top of the stones a membrane is placed, above which 0.5m of excavated material is placed. In most cases, it´s not necessary to build with dry set masonry.

To work out the size, a water infiltration test is done with saturated soil. To do this, a pipe is held vertically on the ground, and half a meter of water is poured in. The time taken for half of the water to be absorbed is measured. This allows the infiltration rate to be calculated (l/s per m^2).

To work out the required surface area, only the vertical walls are included in the calculation. The speed of absorption should be similar to that of the water in the bed. At moments of high flow over a short time (emptying the pipeline) the bed is calculated so that the volume is sufficient to collect all of the water. The shape which has the highest absorption surface area per m^3 of excavation is a narrow trench.

12. 7 VALVE BOXES

A valve box is a structure which protects the valves from unwanted interference. The box should rise above ground level by at least 10cm, to avoid flooding from rainfall. For water from leaks to flow away freely, they have no floor, but instead a layer of gravel. Traditional valve boxes should be large enough to allow maintenance work to take place.

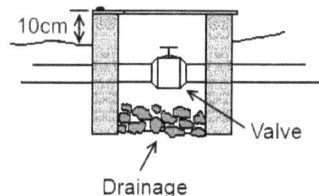

Masonry boxes

Quick and easy to build. Only the upper 40cm requires cement. The rest can be dry stone.

Concrete boxes

Slow and more expensive. They are only really worth building in large numbers using moulds:

Fig. 12.6. Mould for a concrete valve box.

An interesting possibility is using a valve box as a restricting block.

Metal pipe boxes

This is a very interesting alternative for the speed of construction and low cost. The galvanized iron pipe which makes up the box rises above ground level by 10cm and is covered with a threaded end cap. It has 2 grooves which allow it to be slotted over the pipeline. To operate the valve a homemade GI tool is used with two protruding parts. These fit onto the wheel of the valve and allow it to be rotated.

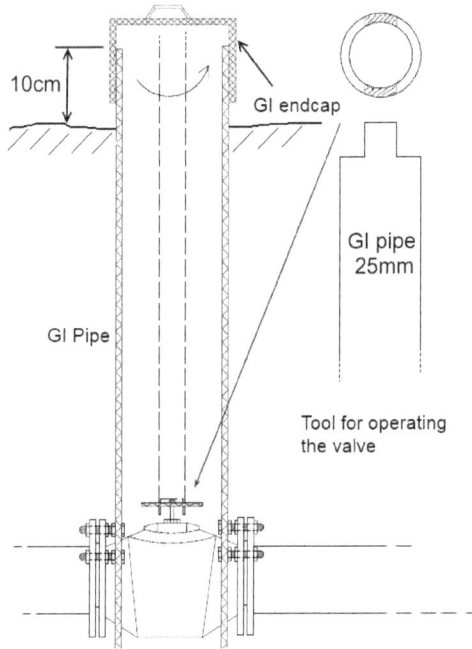

10cm

GI endcap

GI pipe
25mm

GI Pipe

Tool for operating
the valve

12. 8 PUBLIC TAP STANDS

This is the part of the system which is in direct contact with the end users. They are varied and each community will know which kind they want. At the moment of working out the needs, one tap for every 150 people with a minimum flow of 0.25 l/s is planned. Once the tap stand is built, it takes little extra effort and material to add more taps, which can substantially improve the level of service.

Location is important, with distances of less than 300m to the end users. In hot climates, shaded areas are best, and in cold climates, sunny areas. Look for areas with good natural drainage. Waste water can quickly gather in pools and puddles.

Just one tap can create huge pools:

To avoid this, each tap should have its own soakaway. A permeable concrete slab on top of the bed can result in almost no pools or puddles.

The taps are also the most abused parts of the system. They need to be robust:

Bent laterally

Previous repair

One aspect that is often forgotten is the raised platform on which water containers can be rested. Lifting 20kg from a height of 60cm demands is much easier than from the ground, thus making it easier for a child, elderly person, a person with disability from

AIDS, or simply anyone who is doing the same thing 8 or 9 times a day, year after year.
Some models here, far from perfect:

This last model has a platform for the vessel avoiding the effort of lifting from the ground and preventing ground contamination.

13. Documenting the project

Far from exhaustive and in an attempt to provide a few tips and tricks, I thought it´d be useful to go over a few ideas which can come in handy for these kinds of projects.

13. 1 GETTING A SAFE CONDUCT

As with most people, the authorities and partners you will be working with won´t be able to last more than 15 minutes without disagreeing on something. This is a very human trait but can end up causing large delays to a project. The local supervisor of the water board doesn´t agree with a certain way of doing things, the new ACNUR delegate does not approve another...

As soon as you have a detailed design with drawings, make sure you get all those involved to seal and sign a copy of the project. This will avoid problems in the future, when someone chooses to disagree over a certain issue, and can help remind them that they had agreed to it previously.

13. 2 DRAWINGS

It´s rare that what´s in the design drawings is exactly what´s built in real life; unexpected things happen, improvements are thought of...It´s important to include all of these modifications in a new set of drawings, drawings of what has been built in reality, making a clear distinction between design drawings of what was planned and what was actually built.

A third set of drawings are those of the survey, in other words, where the design ideas are thrashed out and where modifications are explained. It´s very important that in each of 3 types of drawings, they are clearly distinguished between each. It´s very frustrating to find the only drawings available of a system full of modifications without knowing if they are simply ideas or what was implemented in the field.

It often happens that all the information for a system is lost all together. When you want to work on it 20 years later, not a single piece of paper can be found. Think of making a black box with all the design drawings and details of what was built in a safe and visible place. For this to work, the box has to be dismantled to access the documents. If it can be easily opened, things will get lost the first time someone needs to find a missing document.

13. 3 NODEBOOKS

A nodebook is a collection of schematics of each point in the system which uses some kind of accessory (node). These points tend to be named consecutively in the direction of flow:

NODE	140						
Product	Material	DN	Connection	Note		Qty	Total
Tee	GI	100*100*100 mm	Triple flange			1	1
Flange adaptor	Metal	100 mm		For PVC class D		2	2
Flange adaptor	Metal	50 mm		For PVC class D		1	1
Reducer	GI	100 to 50 mm	Double flange			1	1
Elbow 45°	PVC class D	100 mm	Double socket			2	2
Gate valve		50 mm	Double flange			1	1
						8	8

NODE	154-165						
Product	Material	DN	Connection	Note		Qty	Total
Tee	GI	150*150*50 mm	Double socket single flange in branch			1	12
Gate valve		50 mm	Double flange			1	12
Flange adaptor	Metal	50 mm	Flange connection	For PVC class D		1	12

Fig. 13.4. Extract from a nodebook, Proja Jadid, Afghanistan.

The nodebook is one of the most useful documents you´ll find. It doesn´t all have to be done in AutoCAD, a simple Excel spreadsheet with a diagram is enough. The diagram can be done quickly but rotating, copying and pasting a set of basic symbols with a simple program such as Paint:

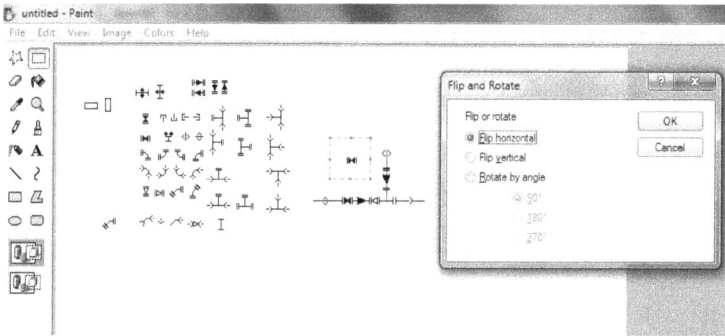

Fig. 13.4b. Putting together a diagram with Paint. Rotating a valve.

You can download the document here: www.arnalich.com/dwnl/simbolos.jpg. The last two icons are used to connect transparently or to delete everything underneath.

13. 4 SIMPLIFIED HYDRAULIC SYMBOLS

Unfortunately, some hydraulic symbols can be very complicated. Luckily this kind of rigour isn´t always necessary, as most node books include a list of components meaning you can identify each thing. I suggest using a simplified and more intuitive version of the symbols, which is nonetheless similar to the rigorous version and easily interpretable to any technical expert.

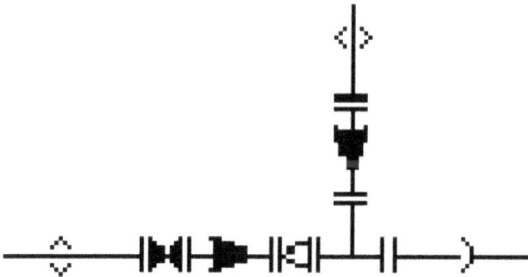

With the symbols on the following page you can cover 95% of accessories.

⊣| Union with bolted flange.

⊣⟨⊢   Glued o elastic PVC union, or threaded GI union

⊩   90° elbow with flange.

    90° glued PVC elbow or threaded GI.

⊤⊢| T with triple flange.

⊥⟨⊢   T with triple glued or threaded union.

‖◄‖   Non-return or check valve (flange). Direction of flow ←.

‖◁‖   Reduction. Very similar to the non-return valve but not filled in.

‖►◄‖   On-off valve (gate, ball…).

⊣⊐   End cap, end of line.

    Coupling or union.

    Air valve.

⊢—⊣   Hose or short length of pipe.

Below you can see what the diagram of a node looks like with these symbols:

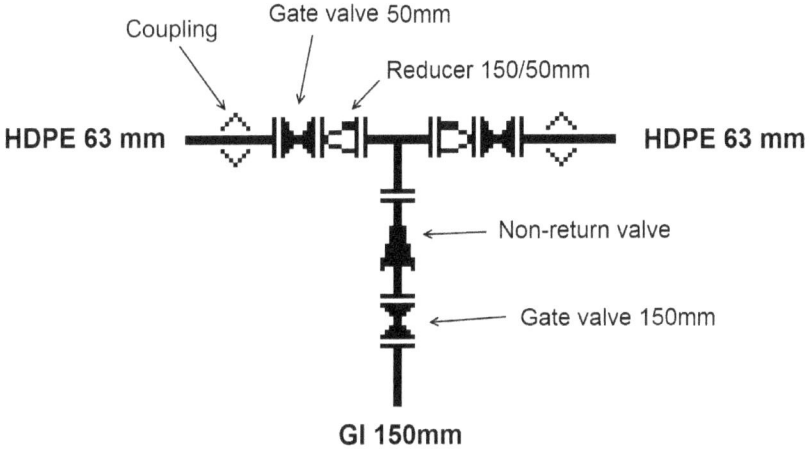

14. Economic aspects

14. 1 INTRODUCTION

If the hydraulic aspects are important for the service to exist, the economic aspects are decisive and can make the difference between the system becoming yet another expensive heap of junk or a key service to power the social and economic development of a community.

The economic evaluation tries to respond to questions such as:
- Which of the different alternatives is cheaper to build?
- Which will be cheaper to operate?
- Can the local population meet the operating costs?
- Will they be able to make the regular payments so that the system does not become obsolete?

Gravity flow water systems tend to have near zero operating costs. However, **they are not completely free to operate and they have an annual repayment cost which is considerable.** That these systems are free is a myth which is often used to justify projects which make no sense. These projects have an initial investment which should be spread over the useful working life of the system. Even if they are completely paid for by donors they have a cost for the local population in terms of the money invested in them for the project which could be used to meet other needs.

Objectives

> ➢ Determine which of the possible alternatives meets the required results at the lowest resource cost.

> ➢ Check that the operating costs of the best alternative are within what the end users are willing to pay, and as such sustainable once the donor has disappeared.

14. 2 COMMON NONSENSE

Aside from the complete absence of an economic evaluation, the 3 most common culprits are:

1. **The "economy of the grandmother":** this has to do with the obsession to save by proposing activities which are really a waste of time and money, compromising the general results and exasperating and de-motivating all those involved. With this focus the most important thing becomes saving pennies wherever possible so that little money is spent.

2. **"Economic despotism":** this has to do with thinking that everything can be worked out economically. To avoid controversy, even where everything is worked out economically, our capacity for measurement is quite limited. For example, what value does education have for a person? Does one dollar have the same value for an average westerner than it does for someone below the poverty line? Nonetheless, that same dollar can buy the same amount of potatoes. What´s wrong here? To be fanatically guided solely by economic efficiency criteria is thoughtless at the best of times because it fails to take into account what is really valued by people.

3. **Vague calculation of diffuse ideas.** This has to do with calculating and budgeting for things that are hardly even defined. To work out what something is going to cost you need to do what that something really is. However many kilometres of pipe, valves, truckloads of sand...working out "how much a water system costs" like this is like working out how much a certain kind of food costs. The design precedes the budget and not vice-versa, as happens so many times in cooperation projects. Statements such as "we have 200,000 dollars, I'd reckon that it will pay for a water system" betray and mess with the results.

14. 3 WILLINGNESS TO PAY

One of the fundamental goals of the economic evaluation is to check that the operating costs of the proposed system is within reach of what the end users are willing to pay, to ensure the project is sustainable once the donor has disappeared.

This means the decision as to whether the investment and operating quotes are acceptable is not down to the designer, but down to the end users. Therefore the person who makes the decisions should be able to communicate well with the future users.

Sometimes the information is close at hand, by observing the cost of traditional systems:

The techniques to work out how much someone would pay go beyond the scope of this book, but a good place to start is the *Willingness-to-pay surveys* published by WEDC, which can be found free on their website here:
http://wedc.lboro.ac.uk/publications/details.php?book=1%2084380%20014%204

14. 4 COSTS

Any activity generates two kinds of costs. The **investment costs** relates to the purchase of the equipment, materials and installation of the system. The **operating costs** refer to the day-to-day running costs.

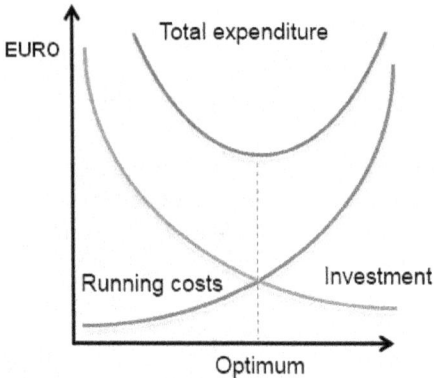

Generally, the greater the initial investment cost, (for example, building tap stands with drainage) the lower the operating costs (chemicals for posterior water treatment). The most economical solution is that which minimises the sum of both costs, the lowest point in the total expenditure curve.

Investment cost

Depends to a great extent on the average projected lifetime of the project. The estimated annual cost is indirect as the value of money changes over time:

- Bank interest. Funds that are used for one thing cannot be invested to make more money. That means money spent on a project cannot at the same time be in a bank earning interest.

- Inflation, or the increasing cost of goods without an associated increase in value. For example, a loaf of bread which cost 5 cents of a Euro when purchased now costs 60. The cost has increased while the value of the good itself remains the same.

To be able to add the costs together and see which is lower, they need to be compared for the same date, generally at the start of the project. The procedure is the following:

1. Find out the **interest, i,** which a bank would offer you if you were to make a deposit of the same amount, and convert it into units of 1. For example, 3% → i = 0.03.

2. Estimate the rate of **inflation, s,** over the period in question. You can look at the rates over certain years in the World Bank data and make a calculation based on that: http://go.worldbank.org/WLW1HK71Q0
   You will have to use your intuition as it´s impossible to know how this will evolve in the future. This will be your parameter, **s,** again, in units of 1.

3. Calculate the **real interest rate, r.** This rate takes into account interest and inflation. If inflation is greater than what a bank will offer, your money is worth more now than it will be in the future. If they are the same, value is maintained, and if the bank´s interest rates are above inflation, the value of your money will increase. This is calculated by:

$$r = \frac{1+i}{1+s} - 1$$

4. Calculate the **repayment rate, a_t,** for T number of years:

$$a_t = \frac{(1+r)^T * r}{(1+r)^T - 1}$$

5. The annual cost of the investment, F, of the sum of M, is:

$$F = M * a_t$$

Calculation example:

A water system budgeted at 100,000 Euros and designed for 30 years is being evaluated in Uzastan, where the banks lend money at 5% and inflation over the past 4 years has been:

Inflation: Uzastan

2001

The interest, i = 0.05 and inflation is estimated at 4.5%, so s = 0.045.

The real interest rate is: $r = \dfrac{1+i}{1+s} - 1 = \dfrac{1+0.05}{1+0.045} - 1 = 0.00478$

The repayment factor is:

$$a_t = \frac{(1+r)^T * r}{(1+r)^T - 1} = \frac{(1+0.00478)^{30} * 0,00478}{(1+0.00478)^{30} - 1} = 0.03586$$

The annual cost is F = 100,000 Euros * 0.03586 year^{-1} = 3586.24 Euros/year, approximately 3586 Euros/year.

Note that this is different to 100,000 Euros/ 30 years = 3333.33 Euros/ year. This is due to the fact that the corrected investment value, called **net present value**, is F * 30 years = 107,587 Euros and not the 100,000 Euros of the investment.

Maintenance costs

Often a gravity flow system has to be compared to a pumped system. In systems which are not gravity fed, the main costs are for pumping, followed by water treatment. Conceptually this cost is much simpler to work out; it´s an inventory of all the costs incurred by the system over one year of operation. Having said that, there are some very evasive costs, such as those incurred from breakages. In cooperation, it´s rare that the economic decisions are going to be so precise, costs from breakages are relatively small in well designed systems, and there are often other criteria which will be imposed with limited margins. My advice is not to take them into account.

Calculation example:

A pumping station fills a tank from which a city is supplied. The pump delivers 10 l/s to the tank and consumes 7.2 kWh according to the manufacturer. The population served is 1000 people and it has been decided that each person will receive 50 liters per day. The Price per kWh is 2 Euros and does not vary over the course of the day.

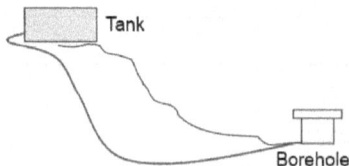

On one hour, the pumping station will pump 10 l/s * 3600 s /h * 1 m³/1000 liters = 36 m³/h.

The hourly cost of pumping will be 7.2 kWh * 2 Euros = 14.4 Euros/hour

The cost per m³, 14.4 Euros/h / 36 m³ /h = 0.4 Eur/m³

Annual demand is: 365 days * 1000 inhab * 50 l/inhab*day * 1 m³/1000 liters = 18250 m³.

And the annual cost will be 18250 m³ * 0.4 Eur/m³ = 7300 Eur.

Economic comparison of alternatives

The procedure is to find out the investment cost and maintenance cost for each alternative and see which sum of both is the cheapest. Below are a few exercises:

Calculation example:

Sharhjaj is situated 12km from the Singag river in India, where the banks lend money at 2%. Two alternatives are being considered:

a.   *A gravity flow project from the river with a total budget of 120,000 Euros. The intake filters the river water and the required chlorine dosage has been established as being 1.7 ppm. The cost of granulated chlorine at 70% is 7 eur/kg.*

b.   *The construction of a project based on a borehole with a total cost of 59,000 Euros. The manufacture's curve (GRUNDFOS) and the pumping condition can be seen below. Electricity is supplied from the town mains and costs 0.2 eur/kWh.*

Which alternative is cheaper for 50,000m³ of water annually?

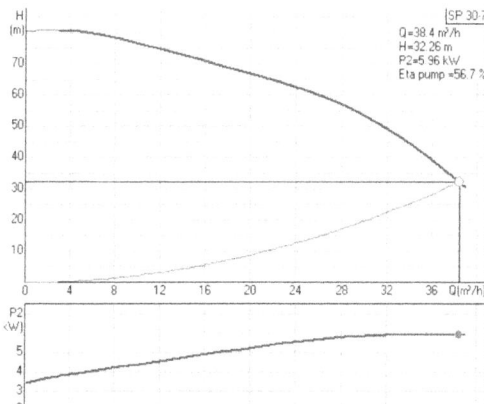

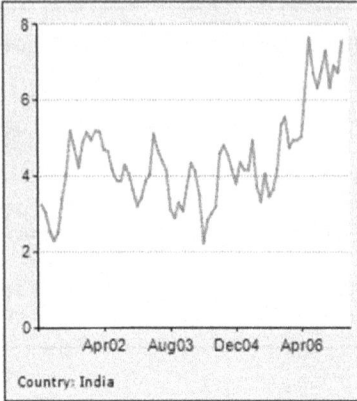

Country: India

Although there is a tendency for increase in the final year, inflation can be said to be around 4%. The annual cost is:

Bank interest i	0.02
Inflation s	0.04
Period (years)	30
Investment M	120000
Real interest rate r	-0.0192
Amortization factor at	0.02432472
Yearly Bill A	**2919**

Bank interest i	0.02
Inflation s	0.04
Period (years)	30
Investment M	59000
Real interest rate r	-0.0192
Amortization factor at	0.02432472
Yearly Bill B	**1435**

Operating costs of alternative A: 1.7 ppm is the same as 1.7 mg/l. Taking into account that chlorine is 70%, the required quantity is:

$$50,000 \ m^3/year * 1.7 \ mg/l * 1,000 \ l/m^3 * 1kg/1,000,000 \ mg \ / \ 0.7$$
$$= 121.43 \ kg/year$$

$$121.43 \ kg/year * 7 \ Eur/kg = \textbf{850 Euros/year}$$

Operating costs of alternative B: the manufacturer's data is 38.4 m^3/h y 5.96 kWh. The number of hours of use and costs are:

$$50,000 \ m^3/ \ year \ / \ 38.4 \ m^3/h = 1302 \ hours/year$$

$$1.302 \ h/year * 5.96 \ kW * 0.2 \ Eur/kWh = \textbf{1552 Euros/year}$$

Option	Gravity	Borehole
Investment	2919	1435
Running costs	850	1552
TOTAL	**3769**	**2987**

Ignoring other criteria, the borehole is the better option.

Calculation example:

Doomborale is situated 6km from the Shabelle river in Ethiopia where the banks lend money at 4%. Again, 2 alternatives are being considered:

    a. The rehabilitation of the canals which arrive close to the settlement, with a total budget of 45,000 Euros.

    b. The construction of a pipeline from the river in a natural depression. The pump is estimated to consume 4,600 Euros per year and the investment is 12,000 Euros.

Which is the cheaper alternative?

According to the World Bank, the evolution of inflation over recent years is the following:

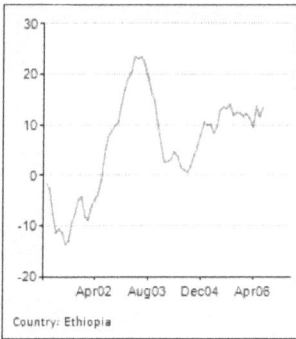

Country: Ethiopia

In countries with variable inflation, prioritise the initial investment over any kind of operating costs. On the one hand, money will quickly lose its value. On the other hand, the cost of goods, notably fuel, will rise far above what the end users can pay. The system will stop working just when the users are most vulnerable.

For this reason, and without making any more calculations, the canals should be rehabilitated.

14. 5 RANKING EXPENSES

Where to spend money on a system and when depends on its nature. Some authors, like Stephenson, offer a generic summary which divides the expenses into: 55% investment in pipes, and 25% for excavation and installation. In cooperation projects, excluding operating NGO expenses and water access (boreholes etc.), the figures below are more appropriate in my experience:

1°. Pipe and accessories	36%
2°. Excavation	31 %
3°. Sand bed	16%
4°. Valve boxes (concrete)	11 %
5°. Pipe installation	5%

Some interesting conclusions can be drawn here. Points 2, 3, 4 and 5 are relatively independent of pipe diameter. That means two thirds of the investment are independent of pipe diameter.

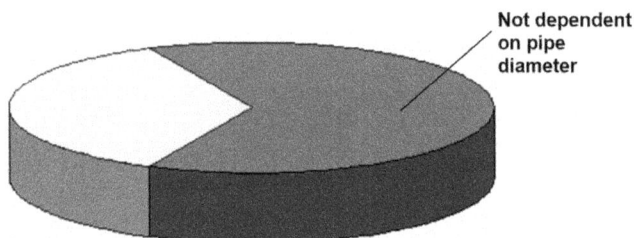

Not dependent on pipe diameter

This conclusion is so important that it merits its own section.

14. 6 DRY *DIAMETERITIS*

This is clearly not an engineering term, but rather a quirk of mine so that you are reminded of one of the most common diseases known to cooperation projects, usually a result of the "economy of the grandmother," or due to the imposition of *this must cost less than X* (so as to be able to present it for that funding). Basically attempts are made to save money on pipe by skimping on diameters and keeping everything to the absolute minimum. The result is a system which has next to no margin for design errors or use changes, little room for enlargement, leaving the users high and dry when they most need water and which are costly to operate. It's no surprise that these systems are often dry as a bone. If they are accompanied by savings in pipe protection material (i.e. sand bed) and excavation costs (buried shallow) the results don't leave much room for celebration.

Doubling up a pipeline because it's no longer able to supply sufficient flow ends up being expensive. In this example, 1000m of 200mm pipe is compared with two pipes of 160mm and 125mm carrying the same amount of water. Note that 200mm is only one pipe diameter up from 160mm.

Option	200mm	160mm+125mm
Cost of pipe	21,000	13,500+8100
Other (64%)	37,300	37,300+37,300
TOTAL (Euros)	58,300	96,200

14. 7 COST vs. DIAMETER

Pipe and carrying capacity

It´s common to think that larger pipes carry more l/s at lower cost. However, this isn´t true for plastic pipe; the cost in relation to carrying capacity remains almost the same for any diameter, a constant. The two graphs below show this for PVC (Uralita) and HDPE (Chresky), around 1 euro per linear meter for every l/s in PVC, and 1.1 Eur/m for HDPE.

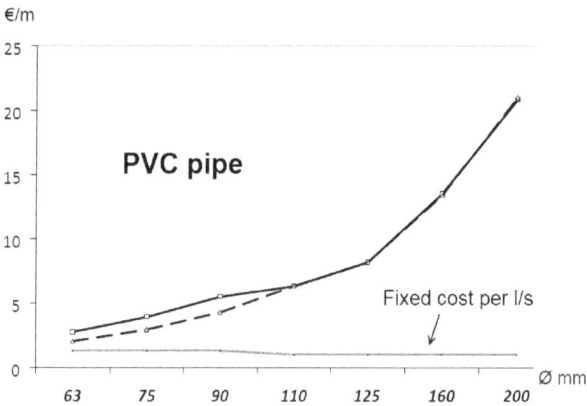

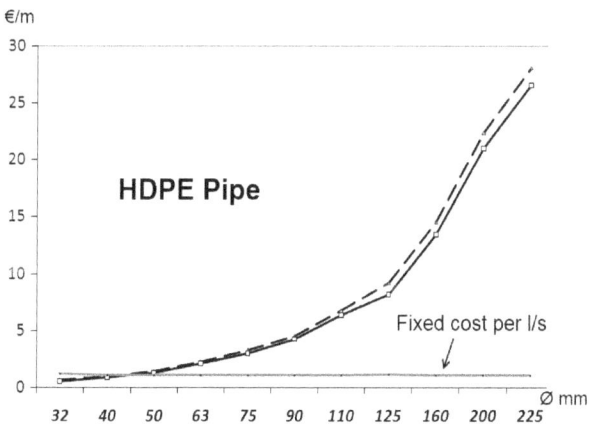

The carrying capacity of the pipe in l/s (continuous) and the cost per meter (discontinuous) increase in parallel as pipe diameter is increased in both cases.

Accessories

An important problem is that of the cost of accessories, especially valves, who´s price rises exponentially with diameter. If a gate valve at 1" costs 11 USD, for 12" it costs 1460 USD. You can see the evolution of cost with diameter in the following graph:

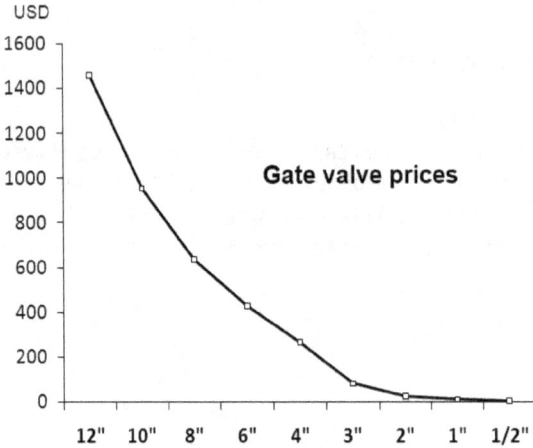

The higher the unit price of an item, the less likely it is that it will be replaced when broken. This may not seem that important, especially when you consider the total investment in the project, but bear in mind that 700 dollars in many places means a large number of working days. The point is, **the replacement of control accessories for larger diameters can represent a considerable hurdle for the community.**

On the surface, the system can continue to function and the local population may not even notice the consequences. Even though they may not be as noticeable as the replacement of a pump (which, when it breaks, means there´s simply no water arriving...) gate valves tend to break when they are being opened or shut, often leaving them half open and reducing the transport capacity of the pipe considerably.

Contracting authorities

One final thought...

When the authorities in charge of water supply have limited human and financial resources, an interesting option is to hire them for the building of the project instead of private companies. With a benefit margin similar to private companies, the project represents income. At the same time, it allows for personnel to be contracted and reinstated, providing on-the-job training and experience in a real project.

This can be a risky option, as you´ll be dealing with a public authority in the case of conflict. It´s very important to include a clause in the contract specifying that the contracting party reserves the right to dissolve the contract at any point, following payment for work done. However, the resulting impact when it works can make the risk worthwhile in many cases.

In the photo, the details on a contract of this type are being discussed with Engineer Patan, of the Kabul Water Authority, who in his day was in charge of 36 full-time engineers and who at the time of the contract had only one.

In remote locations where there are no authorities or organisations in charge of new infrastructure projects, a similar approach can be taken, although less formal in contractual terms, where people are chosen by the community to take up positions in the management of the project. A project with adequate support is a great place to start.

About the author

Santiago Arnalich

At 26 years old, he began as the coordinator of the Kabul Project CAWWS Water Supply, providing water to 565,000 people, probably the most important water supply project to date. Since then, he has designed improvements for more than a million people, including refugee camps in Tanzania, the city of Meulaboh following the Tsunami, and the poor neighbourhoods of Santa Cruz, Bolivia.

Currently he is founder and executive director of Arnalich, Water and Habitat, a private company with a strong social commitment dedicated to promoting the impact of humanitarian organisations through training and technical assistance in the fields of drinking water supply and environmental engineering.

Bibliography

1. Arnalich, S. (2007). *Epanet in Aid: How to calculate water networks by computer.* Arnalich, Water and Habitat

   www.arnalich.com/en/libros.html

2. Arnalich, S. (2007) *Epanet in Aid: 44 Progressive exercises explained step-by-step.* Arnalich, Water and Habitat.

   www.arnalich.com/en/libros.html

3. AWWA (1992). ANSI / AWWA C651-92. Standards for Disinfecting Water Mains.

4. Corcos, G. (2003). *Air in Water Pipes. A manual for designers of spring-supplied gravity-driven drinking water rural delivery systems.* Agua para la vida.

   http://www.aplv.org/Downloads/AirInPipesManual.pdf

5. Davis J. y Lambert R. (2002). *Engineering in Emergencies. A practical guide for relief workers.* 2° Ed. ITDG publishing.

6. Department of Lands, Valuation and Water (1983). *Gravity Fed Rural Piped Water Schemes.* Republic of Malawi.

7. Dipra Technical Committee (1997). *Thrust Restraint Design for Ductile Iron Pipe.* Dipra.

   http://www.dipra.org/pdf/thrustRestraint.pdf

8. Expert Committee, (1999). *Manual on Water Supply and Treatment.* 3° Ed. Government of India.

9. Fernández Navarro, L. (190?). *Investigación y Alumbramiento de Aguas Subterráneas.* Manuales Soler N° 87. Ed. Gallach Calpe.

10. Fuertes, V. S. y otros (2002). *Modelación y Diseño de Redes de Abastecimiento de Agua.* Servicio de Publicación de la Universidad Politécnica de Valencia.

11. Harvey, A. y otros (1993). *Micro-hydro Design Manual. A guide to small-scale water power systems.* Intermediate Technology Publications.

12. Jordan T. D. (1980). *A Handbook of Gravity-Flow Water Systems.* Intermediate Technology Publications.

13. Plastic Pipes Institute (2001). *Disinfection of Newly Constructed Polyethylene Water Mains.*

    http://www.plasticpipe.org/pdf/tr-34_disinfection_for_new_polyethylene_mains.pdf

14. Mays L. W. (1999). *Water Distribution Systems Handbook.* McGraw-Hill Press.

15. Meuli, K. y Wehrle, K. (2001). *Spring Catchment.* Series of Manuals on Drinking Water Supply, Volume 4. SKAT publications.

16. Saint-Gobain Canalisation (2001). *Water Mains Catalogue*.

17. Santosh Kumar Garg (2003). *Water Supply Engineering*. 14º ed. Khanna Publishers.

18. Stephenson, D. (1981). *"Pipeline Design for Water Engineers"*. Ed. Elsevier.

19. Walski, T. M. y otros (2003). *Advanced Water Distribution Modeling and Management*. Haestad Press, USA. Haestad methods.

    http://www.haestad.com/library/books/awdm/online/wwhelp/wwhimpl/js/html/wwhelp.htm

20. Watt, S. B. (1975). *A Manual on the Hydraulic Ram for Pumping Water.* ITDG Publishing.

21. WHO (1996). *Guidelines for drinking-water quality*, 2º **Ed**. Vol. 2 *Health criteria and other supporting information* y *Addendum to Vol. 2* (1998).

    http://www.who.int/water_sanitation_health/dwq/guidelines2/es/index.html (navegar)

APPENDICES

A. PHYSICAL-CHEMICAL DRINKING WATER STANDARDS

Taken from:
Guidelines for Drinking-Water Quality, 2° Ed. Vol. 2 Health criteria and other supporting information, 1996 (pp. 940-949) y Addendum to Vol. 2 1998 (pp. 281-283) Geneva, World Health Organisation.

Detailed data relating to the parameters can be found here:
http://www.who.int/water_sanitation_health/dwq/guidelines2/en/index.html

PHYSICAL STANDARDS:

Parameter		Comments
Salinity	3000 μs/cm	
Turbidity	5 NTU	Removable
pH	<8	For effective chlorination

WITH ADVERSE HEALTH EFFECTS:

Substance	Limit mg/l	Comments
Antimony	0.005	Uncommon; not removable using traditional methods
Arsenic	0.01	Removable
Barium	0.7	Treatment with ionic exchange or precipitation
Boron	0.5	Not removable using traditional methods
Cadmium	0.003	Treatment with precipitation or coagulation
Chromium	0.05	Treatment with coagulation
Copper	2	Uncommon; not removable using traditional methods
Cyanide	0.07	Removable with high dose of chlorine
Fluoride	1.5	Removable with activated aluminium
Lead	0.01	Not present in uncontaminated water
Manganese	0.5	Oxidation (aeration) and filtration
Mercury	0.001	Filtration, sedimentation, ionic exchange…
Molybdenum	0.07	Not removable
Nickel	0.02	Removable using traditional methods
Nitrate (NO_3-)	50	Biological elimination or ionic exchange
Nitrite ($NO2$-)	0.2	Transformation into nitrates through chlorination
Selenium	0.01	Selenium IV with coagulation. Selenium IV not removable
Uranium	0.002	Removable with conventional treatment

WHICH CAN CAUSE COMPLAINTS:

Substance	Limit mg/l	Comments
Aluminium	0.2	Depositions an de-colouring
Copper	1	Clothes and sanitary stains
Iron	0.3	Clothes and sanitary stains
Manganese	0.1	Clothes and sanitary stains
Sodium	200	Bad taste
Sulphates	250	Bad taste, corrosion
Total dissolved solids	1000	Bad taste

BIOLOGICAL:

Parameter		Comments
Coliforms	0	In any 100ml sample

B.   FRICTION LOSS TABLES. PLASTIC PIPE
(Courtesy of Uralita)

Below you will find the friction loss tables for the most commonly used pipes. Due to space limitations not all pipes are listed. If you´re looking for data that isn´t available here, go to www.arnalich.com/dwnl/headloss.zip.

To use the tables you need to know what material you´re working with, the maximum pressure, and the type of water you´ll be transporting (clean/dirty). For a flow of 0.02 l/s, an HDPE of 25mm at 16 bar, carrying clean water (k=0.01), has a head loss of 0.6 m/km.

25 - PN 16	CLEAN WATER: K=0.01	
Head loss (m/km)	Q (l/s)	V (m/s)
0.50	0.018	0.06
0.60	0.020	0.06
0.70	0.022	0.07

J, head loss; Q, flow and V, velocity.

Important: head loss varies somewhat from one manufacturer to another. If a manufacturer provides you with reliable data, use this instead.

Metal pipe is specified with the internal diameter. Plastic pipe with the external diameter. This table shows the approximate internal diameters (ID) for plastic pipe:

ND	25	32	40	50	63	75	90	110	125	140	160	180	200	250	315	400
ID HDPE	20	26	35	44	55	66	79	97	110	123	141	159	176	220	277	353
ID PVC	21	29	36	45	57	68	81	102	115	129	148	159	185	231	291	369

HDPE 25 -ID 20.4mm- PN 16			HDPE 32 - ID 26.2mm- PN 16		
J (m/km)	Q (l/s)	v (m/s)	J (m/km)	Q (l/s)	v (m/s)
0.50	0.018	0.06	0.50	0.037	0.07
0.60	0.020	0.06	0.60	0.041	0.08
0.70	0.022	0.07	0.70	0.045	0.08
0.80	0.024	0.07	0.80	0.049	0.09
0.90	0.026	0.08	0.90	0.052	0.10
1.00	0.028	0.08	1.00	0.056	0.10
1.10	0.029	0.09	1.10	0.059	0.11
1.20	0.031	0.09	1.20	0.062	0.11
1.30	0.032	0.10	1.30	0.065	0.12
1.40	0.034	0.10	1.40	0.068	0.13
1.50	0.035	0.11	1.50	0.071	0.13
1.60	0.037	0.11	1.60	0.073	0.14
1.70	0.038	0.12	1.70	0.076	0.14
1.80	0.039	0.12	1.80	0.079	0.15
1.90	0.041	0.12	1.90	0.081	0.15
2.00	0.042	0.13	2.00	0.084	0.16
2.25	0.045	0.14	2.25	0.090	0.17
2.50	0.048	0.15	2.50	0.095	0.18
2.75	0.050	0.15	2.75	0.101	0.19
3.00	0.053	0.16	3.00	0.106	0.20
3.25	0.056	0.17	3.25	0.111	0.21
3.50	0.058	0.18	3.50	0.116	0.22
3.75	0.061	0.19	3.75	0.121	0.22
4.00	0.063	0.19	4.00	0.125	0.23
4.25	0.065	0.20	4.25	0.130	0.24
4.50	0.067	0.21	4.50	0.134	0.25
4.75	0.070	0.21	4.75	0.139	0.26
5.00	0.072	0.22	5.00	0.143	0.26
5.50	0.076	0.23	5.50	0.151	0.28
6.00	0.080	0.24	6.00	0.159	0.29
6.50	0.084	0.26	6.50	0.166	0.31
7.00	0.087	0.27	7.00	0.173	0.32
7.50	0.091	0.28	7.50	0.180	0.33
8.00	0.094	0.29	8.00	0.187	0.35
8.50	0.098	0.30	8.50	0.194	0.36
9.00	0.101	0.31	9.00	0.200	0.37
10.00	0.107	0.33	10.00	0.213	0.39
12.00	0.119	0.36	12.00	0.236	0.44
15.00	0.136	0.41	15.00	0.269	0.50
20.00	0.160	0.49	20.00	0.316	0.59
30.00	0.202	0.62	30.00	0.398	0.74
45.00	0.254	0.78	45.00	0.501	0.93
60.00	0.299	0.91	60.00	0.589	1.09

HDPE 40 - ID 35.2mm- PN 10			HDPE 40 - ID 32.6mm- PN 16		
J (m/km)	Q (l/s)	v (m/s)	J (m/km)	Q (l/s)	v (m/s)
0.50	0.084	0.09	0.50	0.068	0.08
0.60	0.093	0.10	0.60	0.075	0.09
0.70	0.102	0.10	0.70	0.083	0.10
0.80	0.111	0.11	0.80	0.089	0.11
0.90	0.118	0.12	0.90	0.096	0.11
1.00	0.126	0.13	1.00	0.102	0.12
1.10	0.133	0.14	1.10	0.108	0.13
1.20	0.140	0.14	1.20	0.113	0.14
1.30	0.147	0.15	1.30	0.119	0.14
1.40	0.153	0.16	1.40	0.124	0.15
1.50	0.160	0.16	1.50	0.129	0.15
1.60	0.166	0.17	1.60	0.134	0.16
1.70	0.172	0.18	1.70	0.139	0.17
1.80	0.178	0.18	1.80	0.144	0.17
1.90	0.183	0.19	1.90	0.148	0.18
2.00	0.189	0.19	2.00	0.153	0.18
2.25	0.202	0.21	2.25	0.164	0.20
2.50	0.215	0.22	2.50	0.174	0.21
2.75	0.227	0.23	2.75	0.184	0.22
3.00	0.239	0.25	3.00	0.193	0.23
3.25	0.250	0.26	3.25	0.203	0.24
3.50	0.261	0.27	3.50	0.211	0.25
3.75	0.272	0.28	3.75	0.220	0.26
4.00	0.282	0.29	4.00	0.228	0.27
4.25	0.292	0.30	4.25	0.237	0.28
4.50	0.302	0.31	4.50	0.244	0.29
4.75	0.311	0.32	4.75	0.252	0.30
5.00	0.320	0.33	5.00	0.260	0.31
5.50	0.338	0.35	5.50	0.274	0.33
6.00	0.356	0.37	6.00	0.288	0.35
6.50	0.372	0.38	6.50	0.302	0.36
7.00	0.388	0.40	7.00	0.315	0.38
7.50	0.404	0.42	7.50	0.328	0.39
8.00	0.419	0.43	8.00	0.340	0.41
8.50	0.434	0.45	8.50	0.352	0.42
9.00	0.448	0.46	9.00	0.364	0.44
10.00	0.476	0.49	10.00	0.386	0.46
12.00	0.528	0.54	12.00	0.429	0.51
15.00	0.599	0.62	15.00	0.487	0.58
20.00	0.705	0.72	20.00	0.573	0.69
30.00	0.885	0.91	30.00	0.720	0.86
45.00	1.111	1.14	45.00	0.904	1.08
60.00	1.304	1.34	60.00	1.061	1.27

HDPE 63 - ID 55.4mm- PN 10		
J (m/km)	Q (l/s)	v (m/s)
0.50	0.293	0.12
0.60	0.326	0.14
0.70	0.357	0.15
0.80	0.385	0.16
0.90	0.412	0.17
1.00	0.438	0.18
1.10	0.463	0.19
1.20	0.487	0.20
1.30	0.510	0.21
1.40	0.532	0.22
1.50	0.553	0.23
1.60	0.574	0.24
1.70	0.594	0.25
1.80	0.614	0.25
1.90	0.633	0.26
2.00	0.652	0.27
2.25	0.698	0.29
2.50	0.741	0.31
2.75	0.782	0.32
3.00	0.822	0.34
3.25	0.860	0.36
3.50	0.897	0.37
3.75	0.933	0.39
4.00	0.968	0.40
4.25	1.002	0.42
4.50	1.035	0.43
4.75	1.067	0.44
5.00	1.099	0.46
5.50	1.159	0.48
6.00	1.218	0.51
6.50	1.274	0.53
7.00	1.329	0.55
7.50	1.381	0.57
8.00	1.432	0.59
8.50	1.482	0.61
9.00	1.531	0.63
10.00	1.624	0.67
12.00	1.799	0.75
15.00	2.038	0.85
20.00	2.393	0.99
30.00	2.998	1.24
45.00	3.752	1.56
60.00	4.396	1.82

HDPE 63 - ID 51.4mm- PN 16		
J (m/km)	Q (l/s)	v (m/s)
0.50	0.239	0.12
0.60	0.265	0.13
0.70	0.290	0.14
0.80	0.314	0.15
0.90	0.336	0.16
1.00	0.357	0.17
1.10	0.377	0.18
1.20	0.396	0.19
1.30	0.415	0.20
1.40	0.433	0.21
1.50	0.451	0.22
1.60	0.468	0.23
1.70	0.484	0.23
1.80	0.501	0.24
1.90	0.516	0.25
2.00	0.532	0.26
2.25	0.569	0.27
2.50	0.604	0.29
2.75	0.638	0.31
3.00	0.671	0.32
3.25	0.702	0.34
3.50	0.732	0.35
3.75	0.762	0.37
4.00	0.790	0.38
4.25	0.818	0.39
4.50	0.845	0.41
4.75	0.871	0.42
5.00	0.897	0.43
5.50	0.947	0.46
6.00	0.994	0.48
6.50	1.040	0.50
7.00	1.085	0.52
7.50	1.128	0.54
8.00	1.170	0.56
8.50	1.211	0.58
9.00	1.250	0.60
10.00	1.327	0.64
12.00	1.470	0.71
15.00	1.666	0.80
20.00	1.957	0.94
30.00	2.452	1.18
45.00	3.070	1.48
60.00	3.598	1.73

HDPE 90 - ID 79.2mm- PN 10			HDPE 90 - ID 73.6mm- PN 16		
J (m/km)	Q (l/s)	v (m/s)	J (m/km)	Q (l/s)	v (m/s)
0.50	0.780	0.16	0.50	0.639	0.15
0.60	0.866	0.18	0.60	0.709	0.17
0.70	0.946	0.19	0.70	0.775	0.18
0.80	1.021	0.21	0.80	0.837	0.20
0.90	1.092	0.22	0.90	0.895	0.21
1.00	1.160	0.24	1.00	0.950	0.22
1.10	1.225	0.25	1.10	1.004	0.24
1.20	1.287	0.26	1.20	1.055	0.25
1.30	1.347	0.27	1.30	1.104	0.26
1.40	1.405	0.29	1.40	1.152	0.27
1.50	1.461	0.30	1.50	1.198	0.28
1.60	1.516	0.31	1.60	1.243	0.29
1.70	1.569	0.32	1.70	1.286	0.30
1.80	1.620	0.33	1.80	1.329	0.31
1.90	1.671	0.34	1.90	1.370	0.32
2.00	1.720	0.35	2.00	1.410	0.33
2.25	1.839	0.37	2.25	1.508	0.35
2.50	1.951	0.40	2.50	1.600	0.38
2.75	2.059	0.42	2.75	1.689	0.40
3.00	2.163	0.44	3.00	1.774	0.42
3.25	2.263	0.46	3.25	1.856	0.44
3.50	2.359	0.48	3.50	1.936	0.45
3.75	2.452	0.50	3.75	2.012	0.47
4.00	2.543	0.52	4.00	2.087	0.49
4.25	2.631	0.53	4.25	2.159	0.51
4.50	2.717	0.55	4.50	2.230	0.52
4.75	2.801	0.57	4.75	2.299	0.54
5.00	2.882	0.59	5.00	2.366	0.56
5.50	3.041	0.62	5.50	2.496	0.59
6.00	3.192	0.65	6.00	2.621	0.62
6.50	3.339	0.68	6.50	2.741	0.64
7.00	3.480	0.71	7.00	2.857	0.67
7.50	3.617	0.73	7.50	2.970	0.70
8.00	3.749	0.76	8.00	3.079	0.72
8.50	3.878	0.79	8.50	3.185	0.75
9.00	4.004	0.81	9.00	3.288	0.77
10.00	4.246	0.86	10.00	3.487	0.82
12.00	4.699	0.95	12.00	3.860	0.91
15.00	5.318	1.08	15.00	4.370	1.03
20.00	6.236	1.27	20.00	5.125	1.20
30.00	7.798	1.58	30.00	6.411	1.51
45.00	9.740	1.98	45.00	8.011	1.88
60.00	11.398	2.31	60.00	9.377	2.20

HDPE 110 - ID 96.8mm- PN 10		
J (m/km)	Q (l/s)	v (m/s)
0.50	1.347	0.18
0.60	1.495	0.20
0.70	1.632	0.22
0.80	1.761	0.24
0.90	1.883	0.26
1.00	1.999	0.27
1.10	2.110	0.29
1.20	2.216	0.30
1.30	2.319	0.32
1.40	2.418	0.33
1.50	2.514	0.34
1.60	2.608	0.35
1.70	2.698	0.37
1.80	2.787	0.38
1.90	2.873	0.39
2.00	2.957	0.40
2.25	3.160	0.43
2.50	3.353	0.46
2.75	3.537	0.48
3.00	3.714	0.50
3.25	3.885	0.53
3.50	4.049	0.55
3.75	4.209	0.57
4.00	4.364	0.59
4.25	4.514	0.61
4.50	4.661	0.63
4.75	4.804	0.65
5.00	4.943	0.67
5.50	5.213	0.71
6.00	5.472	0.74
6.50	5.722	0.78
7.00	5.962	0.81
7.50	6.196	0.84
8.00	6.422	0.87
8.50	6.642	0.90
9.00	6.856	0.93
10.00	7.268	0.99
12.00	8.040	1.09
15.00	9.095	1.24
20.00	10.656	1.45
30.00	13.312	1.81
45.00	16.612	2.26
60.00	19.426	2.64

HDPE 110 - ID 90mm- PN 16		
J (m/km)	Q (l/s)	v (m/s)
0.50	1.105	0.17
0.60	1.227	0.19
0.70	1.339	0.21
0.80	1.445	0.23
0.90	1.546	0.24
1.00	1.641	0.26
1.10	1.732	0.27
1.20	1.820	0.29
1.30	1.904	0.30
1.40	1.986	0.31
1.50	2.065	0.32
1.60	2.142	0.34
1.70	2.217	0.35
1.80	2.289	0.36
1.90	2.360	0.37
2.00	2.430	0.38
2.25	2.597	0.41
2.50	2.755	0.43
2.75	2.907	0.46
3.00	3.053	0.48
3.25	3.193	0.50
3.50	3.329	0.52
3.75	3.460	0.54
4.00	3.588	0.56
4.25	3.712	0.58
4.50	3.832	0.60
4.75	3.950	0.62
5.00	4.065	0.64
5.50	4.287	0.67
6.00	4.501	0.71
6.50	4.706	0.74
7.00	4.905	0.77
7.50	5.097	0.80
8.00	5.283	0.83
8.50	5.464	0.86
9.00	5.641	0.89
10.00	5.981	0.94
12.00	6.617	1.04
15.00	7.486	1.18
20.00	8.774	1.38
30.00	10.965	1.72
45.00	13.688	2.15
60.00	16.010	2.52

HDPE 160 - ID 141mm- PN 10		
J (m/km)	Q (l/s)	v (m/s)
0.50	3.732	0.24
0.60	4.136	0.26
0.70	4.512	0.29
0.80	4.865	0.31
0.90	5.198	0.33
1.00	5.515	0.35
1.10	5.818	0.37
1.20	6.110	0.39
1.30	6.390	0.41
1.40	6.661	0.43
1.50	6.923	0.44
1.60	7.178	0.46
1.70	7.426	0.48
1.80	7.667	0.49
1.90	7.902	0.51
2.00	8.131	0.52
2.25	8.684	0.56
2.50	9.209	0.59
2.75	9.711	0.62
3.00	10.193	0.65
3.25	10.656	0.68
3.50	11.104	0.71
3.75	11.538	0.74
4.00	11.959	0.77
4.25	12.367	0.79
4.50	12.765	0.82
4.75	13.154	0.84
5.00	13.532	0.87
5.50	14.265	0.91
6.00	14.968	0.96
6.50	15.644	1.00
7.00	16.298	1.04
7.50	16.930	1.08
8.00	17.543	1.12
8.50	18.138	1.16
9.00	18.718	1.20
10.00	19.835	1.27
12.00	21.924	1.40
15.00	24.775	1.59
20.00	28.994	1.86
30.00	36.156	2.32
45.00	45.043	2.88
60.00	52.609	3.37

HDPE 160 - ID 130.8mm- PN 16		
J (m/km)	Q (l/s)	v (m/s)
0.50	3.046	0.23
0.60	3.377	0.25
0.70	3.685	0.27
0.80	3.973	0.30
0.90	4.246	0.32
1.00	4.506	0.34
1.10	4.754	0.35
1.20	4.992	0.37
1.30	5.222	0.39
1.40	5.443	0.41
1.50	5.658	0.42
1.60	5.867	0.44
1.70	6.069	0.45
1.80	6.267	0.47
1.90	6.459	0.48
2.00	6.647	0.49
2.25	7.100	0.53
2.50	7.530	0.56
2.75	7.941	0.59
3.00	8.335	0.62
3.25	8.715	0.65
3.50	9.082	0.68
3.75	9.438	0.70
4.00	9.782	0.73
4.25	10.117	0.75
4.50	10.443	0.78
4.75	10.761	0.80
5.00	11.072	0.82
5.50	11.672	0.87
6.00	12.248	0.91
6.50	12.803	0.95
7.00	13.338	0.99
7.50	13.856	1.03
8.00	14.359	1.07
8.50	14.847	1.10
9.00	15.322	1.14
10.00	16.238	1.21
12.00	17.951	1.34
15.00	20.290	1.51
20.00	23.750	1.77
30.00	29.627	2.20
45.00	36.921	2.75
60.00	43.133	3.21

HDPE 200 - ID 176.2mm- PN 10			HDPE 200 - ID 163.6mm- PN 16		
J (m/km)	Q (l/s)	v (m/s)	J (m/km)	Q (l/s)	v (m/s)
0.50	6.805	0.28	0.50	5.573	0.27
0.60	7.539	0.31	0.60	6.175	0.29
0.70	8.221	0.34	0.70	6.734	0.32
0.80	8.860	0.36	0.80	7.258	0.35
0.90	9.463	0.39	0.90	7.754	0.37
1.00	10.038	0.41	1.00	8.225	0.39
1.10	10.587	0.43	1.10	8.676	0.41
1.20	11.114	0.46	1.20	9.108	0.43
1.30	11.621	0.48	1.30	9.525	0.45
1.40	12.112	0.50	1.40	9.928	0.47
1.50	12.586	0.52	1.50	10.317	0.49
1.60	13.047	0.54	1.60	10.695	0.51
1.70	13.495	0.55	1.70	11.063	0.53
1.80	13.931	0.57	1.80	11.421	0.54
1.90	14.356	0.59	1.90	11.770	0.56
2.00	14.771	0.61	2.00	12.111	0.58
2.25	15.769	0.65	2.25	12.931	0.62
2.50	16.719	0.69	2.50	13.710	0.65
2.75	17.625	0.72	2.75	14.455	0.69
3.00	18.496	0.76	3.00	15.170	0.72
3.25	19.333	0.79	3.25	15.858	0.75
3.50	20.142	0.83	3.50	16.523	0.79
3.75	20.925	0.86	3.75	17.166	0.82
4.00	21.684	0.89	4.00	17.790	0.85
4.25	22.422	0.92	4.25	18.396	0.88
4.50	23.140	0.95	4.50	18.986	0.90
4.75	23.840	0.98	4.75	19.562	0.93
5.00	24.524	1.01	5.00	20.123	0.96
5.50	25.846	1.06	5.50	21.209	1.01
6.00	27.113	1.11	6.00	22.251	1.06
6.50	28.333	1.16	6.50	23.254	1.11
7.00	29.510	1.21	7.00	24.221	1.15
7.50	30.650	1.26	7.50	25.158	1.20
8.00	31.754	1.30	8.00	26.066	1.24
8.50	32.828	1.35	8.50	26.948	1.28
9.00	33.872	1.39	9.00	27.807	1.32
10.00	35.884	1.47	10.00	29.461	1.40
12.00	39.645	1.63	12.00	32.554	1.55
15.00	44.778	1.84	15.00	36.775	1.75
20.00	52.366	2.15	20.00	43.017	2.05
30.00	65.239	2.68	30.00	53.609	2.55
45.00	81.195	3.33	45.00	66.741	3.17
60.00	94.771	3.89	60.00	77.918	3.71

PVC 40 - ID 36.2mm- PN 10			PVC 40 - ID 34mm- PN 16		
J (m/km)	Q (l/s)	v m/s)	J (m/km)	Q (l/s)	v (m/s)
0.50	0.091	0.09	0.50	0.076	0.08
0.60	0.101	0.10	0.60	0.085	0.09
0.70	0.110	0.11	0.70	0.093	0.10
0.80	0.119	0.12	0.80	0.100	0.11
0.90	0.128	0.12	0.90	0.108	0.12
1.00	0.136	0.13	1.00	0.114	0.13
1.10	0.144	0.14	1.10	0.121	0.13
1.20	0.151	0.15	1.20	0.127	0.14
1.30	0.159	0.15	1.30	0.133	0.15
1.40	0.166	0.16	1.40	0.139	0.15
1.50	0.172	0.17	1.50	0.145	0.16
1.60	0.179	0.17	1.60	0.151	0.17
1.70	0.186	0.18	1.70	0.156	0.17
1.80	0.192	0.19	1.80	0.161	0.18
1.90	0.198	0.19	1.90	0.167	0.18
2.00	0.204	0.20	2.00	0.172	0.19
2.25	0.218	0.21	2.25	0.184	0.20
2.50	0.232	0.23	2.50	0.195	0.22
2.75	0.245	0.24	2.75	0.206	0.23
3.00	0.258	0.25	3.00	0.217	0.24
3.25	0.270	0.26	3.25	0.227	0.25
3.50	0.282	0.27	3.50	0.237	0.26
3.75	0.293	0.28	3.75	0.247	0.27
4.00	0.304	0.30	4.00	0.256	0.28
4.25	0.315	0.31	4.25	0.265	0.29
4.50	0.326	0.32	4.50	0.274	0.30
4.75	0.336	0.33	4.75	0.283	0.31
5.00	0.346	0.34	5.00	0.291	0.32
5.50	0.365	0.35	5.50	0.308	0.34
6.00	0.384	0.37	6.00	0.324	0.36
6.50	0.402	0.39	6.50	0.339	0.37
7.00	0.419	0.41	7.00	0.353	0.39
7.50	0.436	0.42	7.50	0.368	0.40
8.00	0.452	0.44	8.00	0.381	0.42
8.50	0.468	0.45	8.50	0.395	0.43
9.00	0.484	0.47	9.00	0.408	0.45
10.00	0.514	0.50	10.00	0.433	0.48
12.00	0.570	0.55	12.00	0.481	0.53
15.00	0.646	0.63	15.00	0.545	0.60
20.00	0.760	0.74	20.00	0.642	0.71
30.00	0.955	0.93	30.00	0.806	0.89
45.00	1.198	1.16	45.00	1.012	1.11
60.00	1.406	1.37	60.00	1.188	1.31

PVC 63 - ID 57mm- PN 10			PVC 63 - ID 53.6mm- PN 16		
J (m/km)	Q (l/s)	v m/s)	J (m/km)	Q (l/s)	v (m/s)
0.50	0.317	0.12	0.50	0.268	0.12
0.60	0.353	0.14	0.60	0.298	0.13
0.70	0.385	0.15	0.70	0.326	0.14
0.80	0.416	0.16	0.80	0.352	0.16
0.90	0.446	0.17	0.90	0.377	0.17
1.00	0.474	0.19	1.00	0.400	0.18
1.10	0.500	0.20	1.10	0.423	0.19
1.20	0.526	0.21	1.20	0.445	0.20
1.30	0.551	0.22	1.30	0.466	0.21
1.40	0.575	0.23	1.40	0.486	0.22
1.50	0.598	0.23	1.50	0.506	0.22
1.60	0.620	0.24	1.60	0.525	0.23
1.70	0.642	0.25	1.70	0.543	0.24
1.80	0.664	0.26	1.80	0.561	0.25
1.90	0.684	0.27	1.90	0.579	0.26
2.00	0.705	0.28	2.00	0.596	0.26
2.25	0.754	0.30	2.25	0.638	0.28
2.50	0.801	0.31	2.50	0.677	0.30
2.75	0.845	0.33	2.75	0.715	0.32
3.00	0.888	0.35	3.00	0.752	0.33
3.25	0.929	0.36	3.25	0.787	0.35
3.50	0.969	0.38	3.50	0.820	0.36
3.75	1.008	0.40	3.75	0.853	0.38
4.00	1.046	0.41	4.00	0.885	0.39
4.25	1.082	0.42	4.25	0.916	0.41
4.50	1.118	0.44	4.50	0.946	0.42
4.75	1.153	0.45	4.75	0.976	0.43
5.00	1.187	0.47	5.00	1.005	0.45
5.50	1.252	0.49	5.50	1.060	0.47
6.00	1.315	0.52	6.00	1.114	0.49
6.50	1.376	0.54	6.50	1.165	0.52
7.00	1.435	0.56	7.00	1.215	0.54
7.50	1.492	0.58	7.50	1.263	0.56
8.00	1.547	0.61	8.00	1.310	0.58
8.50	1.601	0.63	8.50	1.356	0.60
9.00	1.653	0.65	9.00	1.400	0.62
10.00	1.754	0.69	10.00	1.486	0.66
12.00	1.942	0.76	12.00	1.646	0.73
15.00	2.200	0.86	15.00	1.865	0.83
20.00	2.583	1.01	20.00	2.190	0.97
30.00	3.236	1.27	30.00	2.744	1.22
45.00	4.049	1.59	45.00	3.435	1.52
60.00	4.744	1.86	60.00	4.025	1.78

PVC 90 - ID 81.4mm- PN 10		
J (m/km)	Q (l/s)	v (m/s)
0.50	0.841	0.16
0.60	0.933	0.18
0.70	1.019	0.20
0.80	1.100	0.21
0.90	1.177	0.23
1.00	1.250	0.24
1.10	1.319	0.25
1.20	1.386	0.27
1.30	1.451	0.28
1.40	1.513	0.29
1.50	1.574	0.30
1.60	1.632	0.31
1.70	1.689	0.32
1.80	1.745	0.34
1.90	1.799	0.35
2.00	1.852	0.36
2.25	1.980	0.38
2.50	2.101	0.40
2.75	2.217	0.43
3.00	2.329	0.45
3.25	2.436	0.47
3.50	2.540	0.49
3.75	2.640	0.51
4.00	2.738	0.53
4.25	2.833	0.54
4.50	2.925	0.56
4.75	3.015	0.58
5.00	3.103	0.60
5.50	3.273	0.63
6.00	3.436	0.66
6.50	3.594	0.69
7.00	3.746	0.72
7.50	3.893	0.75
8.00	4.035	0.78
8.50	4.174	0.80
9.00	4.309	0.83
10.00	4.570	0.88
12.00	5.057	0.97
15.00	5.723	1.10
20.00	6.710	1.29
30.00	8.389	1.61
45.00	10.478	2.01
60.00	12.260	2.36

PVC 90 - ID 76.6mm- PN 16		
J (m/km)	Q (l/s)	v (m/s)
0.50	0.712	0.15
0.60	0.791	0.17
0.70	0.864	0.19
0.80	0.933	0.20
0.90	0.998	0.22
1.00	1.060	0.23
1.10	1.119	0.24
1.20	1.176	0.26
1.30	1.230	0.27
1.40	1.283	0.28
1.50	1.335	0.29
1.60	1.385	0.30
1.70	1.433	0.31
1.80	1.480	0.32
1.90	1.526	0.33
2.00	1.571	0.34
2.25	1.680	0.36
2.50	1.783	0.39
2.75	1.882	0.41
3.00	1.976	0.43
3.25	2.068	0.45
3.50	2.156	0.47
3.75	2.241	0.49
4.00	2.324	0.50
4.25	2.405	0.52
4.50	2.483	0.54
4.75	2.560	0.56
5.00	2.635	0.57
5.50	2.779	0.60
6.00	2.918	0.63
6.50	3.052	0.66
7.00	3.181	0.69
7.50	3.306	0.72
8.00	3.428	0.74
8.50	3.546	0.77
9.00	3.661	0.79
10.00	3.882	0.84
12.00	4.297	0.93
15.00	4.864	1.06
20.00	5.704	1.24
30.00	7.133	1.55
45.00	8.912	1.93
60.00	10.429	2.26

PVC 110 - ID 101.6mm- PN 10		
J (m/km)	Q (l/s)	v (m/s)
0.50	1.536	0.19
0.60	1.705	0.21
0.70	1.861	0.23
0.80	2.008	0.25
0.90	2.146	0.26
1.00	2.278	0.28
1.10	2.405	0.30
1.20	2.526	0.31
1.30	2.643	0.33
1.40	2.756	0.34
1.50	2.865	0.35
1.60	2.971	0.37
1.70	3.075	0.38
1.80	3.175	0.39
1.90	3.273	0.40
2.00	3.369	0.42
2.25	3.600	0.44
2.50	3.819	0.47
2.75	4.029	0.50
3.00	4.231	0.52
3.25	4.425	0.55
3.50	4.612	0.57
3.75	4.793	0.59
4.00	4.970	0.61
4.25	5.141	0.63
4.50	5.307	0.65
4.75	5.470	0.67
5.00	5.629	0.69
5.50	5.936	0.73
6.00	6.230	0.77
6.50	6.514	0.80
7.00	6.788	0.84
7.50	7.053	0.87
8.00	7.310	0.90
8.50	7.560	0.93
9.00	7.803	0.96
10.00	8.272	1.02
12.00	9.150	1.13
15.00	10.349	1.28
20.00	12.124	1.50
30.00	15.142	1.87
45.00	18.891	2.33
60.00	22.088	2.72

PVC 110 - ID 96.8mm- PN 16		
J (m/km)	Q (l/s)	v (m/s)
0.50	1.347	0.18
0.60	1.495	0.20
0.70	1.632	0.22
0.80	1.761	0.24
0.90	1.883	0.26
1.00	1.999	0.27
1.10	2.110	0.29
1.20	2.216	0.30
1.30	2.319	0.32
1.40	2.418	0.33
1.50	2.514	0.34
1.60	2.608	0.35
1.70	2.698	0.37
1.80	2.787	0.38
1.90	2.873	0.39
2.00	2.957	0.40
2.25	3.160	0.43
2.50	3.353	0.46
2.75	3.537	0.48
3.00	3.714	0.50
3.25	3.885	0.53
3.50	4.049	0.55
3.75	4.209	0.57
4.00	4.364	0.59
4.25	4.514	0.61
4.50	4.661	0.63
4.75	4.804	0.65
5.00	4.943	0.67
5.50	5.213	0.71
6.00	5.472	0.74
6.50	5.722	0.78
7.00	5.962	0.81
7.50	6.196	0.84
8.00	6.422	0.87
8.50	6.642	0.90
9.00	6.856	0.93
10.00	7.268	0.99
12.00	8.040	1.09
15.00	9.095	1.24
20.00	10.656	1.45
30.00	13.312	1.81
45.00	16.612	2.26
60.00	19.426	2.64

PVC 160 - ID 147.6mm- PN 10			PVC 160 - ID 141mm- PN 16		
J (m/km)	Q (l/s)	v (m/s)	J (m/km)	Q (l/s)	v (m/s)
0.50	4.222	0.25	0.50	3.732	0.24
0.60	4.680	0.27	0.60	4.136	0.26
0.70	5.104	0.30	0.70	4.512	0.29
0.80	5.503	0.32	0.80	4.865	0.31
0.90	5.879	0.34	0.90	5.198	0.33
1.00	6.238	0.36	1.00	5.515	0.35
1.10	6.580	0.38	1.10	5.818	0.37
1.20	6.909	0.40	1.20	6.110	0.39
1.30	7.226	0.42	1.30	6.390	0.41
1.40	7.532	0.44	1.40	6.661	0.43
1.50	7.828	0.46	1.50	6.923	0.44
1.60	8.116	0.47	1.60	7.178	0.46
1.70	8.395	0.49	1.70	7.426	0.48
1.80	8.668	0.51	1.80	7.667	0.49
1.90	8.933	0.52	1.90	7.902	0.51
2.00	9.193	0.54	2.00	8.131	0.52
2.25	9.816	0.57	2.25	8.684	0.56
2.50	10.409	0.61	2.50	9.209	0.59
2.75	10.976	0.64	2.75	9.711	0.62
3.00	11.520	0.67	3.00	10.193	0.65
3.25	12.044	0.70	3.25	10.656	0.68
3.50	12.550	0.73	3.50	11.104	0.71
3.75	13.039	0.76	3.75	11.538	0.74
4.00	13.514	0.79	4.00	11.959	0.77
4.25	13.976	0.82	4.25	12.367	0.79
4.50	14.425	0.84	4.50	12.765	0.82
4.75	14.863	0.87	4.75	13.154	0.84
5.00	15.291	0.89	5.00	13.532	0.87
5.50	16.118	0.94	5.50	14.265	0.91
6.00	16.911	0.99	6.00	14.968	0.96
6.50	17.675	1.03	6.50	15.644	1.00
7.00	18.412	1.08	7.00	16.298	1.04
7.50	19.125	1.12	7.50	16.930	1.08
8.00	19.817	1.16	8.00	17.543	1.12
8.50	20.490	1.20	8.50	18.138	1.16
9.00	21.144	1.24	9.00	18.718	1.20
10.00	22.404	1.31	10.00	19.835	1.27
12.00	24.761	1.45	12.00	21.924	1.40
15.00	27.979	1.64	15.00	24.775	1.59
20.00	32.738	1.91	20.00	28.994	1.86
30.00	40.818	2.39	30.00	36.156	2.32
45.00	50.839	2.97	45.00	45.043	2.88
60.00	59.371	3.47	60.00	52.609	3.37

J (m/km)	Q (l/s)	v (m/s)
0.50	7.714	0.29
0.60	8.545	0.32
0.70	9.316	0.35
0.80	10.04	0.38
0.90	10.723	0.4
1.00	11.373	0.42
1.10	11.995	0.45
1.20	12.591	0.47
1.30	13.166	0.49
1.40	13.721	0.51
1.50	14.258	0.53
1.60	14.779	0.55
1.70	15.286	0.57
1.80	15.779	0.59
1.90	16.26	0.61
2.00	16.73	0.63
2.25	17.86	0.67
2.50	18.934	0.71
2.75	19.96	0.75
3.00	20.944	0.78
3.25	21.892	0.82
3.50	22.807	0.85
3.75	23.692	0.89
4.00	24.551	0.92
4.25	25.386	0.95
4.50	26.198	0.98
4.75	26.99	1.01
5.00	27.763	1.04
5.50	29.258	1.09
6.00	30.691	1.15
6.50	32.071	1.2
7.00	33.402	1.25
7.50	34.691	1.3
8.00	35.94	1.34
8.50	37.154	1.39
9.00	38.335	1.43
10.00	40.609	1.52
12.00	44.862	1.68
15.00	50.665	1.89
20.00	59.242	2.21
30.00	73.791	2.76
45.00	91.821	3.43
60.00	107.159	4.00

PVC 200 - ID 184.6mm- PN 10

PVC 200 - ID 176.2mm- PN 16

J (m/km)	Q (l/s)	v (m/s)
0.50	6.805	0.28
0.60	7.539	0.31
0.70	8.221	0.34
0.80	8.86	0.36
0.90	9.463	0.39
1.00	10.038	0.41
1.10	10.587	0.43
1.20	11.114	0.46
1.30	11.621	0.48
1.40	12.112	0.5
1.50	12.586	0.52
1.60	13.047	0.54
1.70	13.495	0.55
1.80	13.931	0.57
1.90	14.356	0.59
2.00	14.771	0.61
2.25	15.769	0.65
2.50	16.719	0.69
2.75	17.625	0.72
3.00	18.496	0.76
3.25	19.333	0.79
3.50	20.142	0.83
3.75	20.925	0.86
4.00	21.684	0.89
4.25	22.422	0.92
4.50	23.14	0.95
4.75	23.84	0.98
5.00	24.524	1.01
5.50	25.846	1.06
6.00	27.113	1.11
6.50	28.333	1.16
7.00	29.51	1.21
7.50	30.65	1.26
8.00	31.754	1.3
8.50	32.828	1.35
9.00	33.872	1.39
10.00	35.88	1.47
12.00	39.65	1.63
15.00	44.78	1.84
20.00	52.37	2.15
30.00	65.24	2.68
45.00	81.20	3.33
60.00	94.77	3.89

C. FRICTION LOSS TABLE. GALVANIZED IRON

Approximate values for head loss in m/km of galvanized iron pipe, calculated using the Hazen-Williams formula for middle-aged pipe.

Important: head loss varies somewhat from one manufacturer to another. If a manufacturer provides you with reliable data, use this instead.

Flow	1/2"	1"	1 1/2"	2"	3"	4"	5"	6"
l/s	15mm	25mm	40mm	50mm	80mm	100mm	125mm	150mm
0.02	2.28							
0.05	12.46	1.22						
0.1	45.00	4.39						
0.15	95.34	9.31	0.94					
0.2	162.44	15.86	1.61					
0.25	245.56	23.97	2.43					
0.3	344.19	33.60	3.41	1.15				
0.35	457.92	44.71	4.53	1.53				
0.4	586.40	57.25	5.80	1.96				
0.45	729.33	71.20	7.22	2.43				
0.5	886.48	86.55	8.77	2.96				
0.6		121.31	12.30	4.15				
0.7		161.39	16.36	5.52				
0.8		206.67	20.95	7.07				
0.9		257.05	26.06	8.79				
1		312.43	31.67	10.68	0.92			
1.1		372.75	37.79	12.75	1.10			
1.2		437.93	44.40	14.98	1.29			
1.3		507.90	51.49	17.37	1.50			
1.4		582.62	59.06	19.92	1.72			
1.5		662.03	67.11	22.64	1.95			
1.6		746.08	75.63	25.51	2.20			
1.7			84.62	28.55	2.46			
1.8			94.07	31.73	2.74			
1.9			103.98	35.07	3.03	1.02		
2			57.13	38.57	3.33	1.12		
2.2			64.38	46.02	3.97	1.34		

Head loss values in m/km

Values calculated without velocity adjustments (valid for velocities of less than 3 m/s). Coefficient C-110 used for pipe of less than 3" diameter, and C-120 for 3" and over, for middle aged pipe with neutral water (Langelier Index value of ± 0.5).

l/s	15mm	25mm	40mm	50mm	80mm	100mm	125mm	150mm
2.4			72.03	54.06	4.66	1.57		
2.6			80.07	62.70	5.41	1.83		
2.8			88.50	71.92	6.21	2.09		
3			97.32	81.73	7.05	2.38		
3.2			116.11	92.10	7.95	2.68		
3.4			136.41	103.05	8.89	3.00	1.01	
3.6			158.21	114.55	9.88	3.33	1.12	
3.8			181.49	126.62	10.93	3.69	1.24	
4			206.22	139.24	12.01	4.05	1.37	
4.5			232.41	173.18	14.94	5.04	1.70	
5				210.49	18.16	6.13	2.07	
5.5				251.13	21.67	7.31	2.47	1.01
6				295.04	25.46	8.59	2.90	1.19
6.5				342.18	29.53	9.96	3.36	1.38
7					33.87	11.43	3.85	1.59
8					43.37	14.63	4.94	2.03
9					53.94	18.20	6.14	2.53
10					65.57	22.12	7.46	3.07
11					78.23	26.39	8.90	3.66
12					91.90	31.00	10.46	4.30
15					138.94	46.87	15.81	6.51
20					236.70	79.84	26.93	11.08
25					357.83	120.70	40.72	16.76
30						169.19	57.07	23.49
40						288.24	97.23	40.01
50						435.75	146.99	60.49
l/s	15mm	25mm	40mm	50mm	80mm	100mm	125mm	150mm
Flow	1/2"	1"	1 1/2"	2"	3"	4"	5"	6"

To calculate intermediate values, you can use the Hazen-Williams formula, taking into account the warnings and values detailed in the box:

$$h = \frac{10,7LQ^{1,852}}{C^{1,852}D^{4,87}}$$

Where: h, head loss in meters; L, length in meters; C, friction coeffcient and D, diamater in meters and Q flow in m^3/s.

Arnalich. Water and habitat   www.arnalich.com

D.   HDPE PRICE AND LOGISTICS

Prices are approximate from the year 2007, not including transport and taxes. The price and length by roll are included for reference purposes and can change from one supplier to the next.

PN 10 (SDR 11)				
Diameter	Weight	Roll weight	Roll length	Gross price
mm	kg/m	kg	m	EURO/m
25	0.21	128	250	0.501
32	0.27	129	200	0.629
40	0.42	138	150	0.918
50	0.66	141	100	1.445
63	1.04	179	100	2.242
75	1.47	222	100	3.242
90	2.11	286	100	4.518
110	3.14	389	100	6.757
125	4.06	----	----	9.213
160	6.67	----	----	14.518
200	10.4	----	----	22.400
225	13.1	----	----	28.108

The rolls have a diameter of approximately 1.5m, and support weights of around 75kg. 90mm and 110mm rolls often come in lengths of only 25m.

E.   PVC PRICE AND LOGISTICS

Prices are approximate from the year 2007, not including transport and taxes.

Diameter	Weight	Pipe weight	Pipe length	Gross price
mm	kg/m	kg	m	EURO/m
63	0.85	5	6	2.76
75	1.25	7	6	3.94
90	1.90	11	6	5.52
110	3.16	19	6	6.36
125	4.74	28	6	8.19
160	6.84	41	6	13.52
200	10.46	63	6	20.86
250	16.23	97	6	32.46
315	22.81	137	6	51.53
400	----	----	6	82.54

D Class elastic union

Diameter	Weight	Pipe weight	Pipe length	Gross price
mm	kg/m	kg/m	m	EURO/m
40	0.42	3	6	1.18
50	0.59	4	6	1.73
63	0.82	5	6	2.66
75	1.2	7	6	3.73
90	1.82	11	6	5.34
110	3.03	18	6	6.14
125	4.55	27	6	7.92
160	6.57	39	6	13.09
200	10.05	60	6	20.19
250	15.59	94	6	31.42
315	21.91	131	6	49.88

D Class glued union

F.   BUTT FUSION WELDING PROCESS FOR HDPE

The exact times depend on the welding machine you are using. Some automatically apply the required pressure, although due to their cost it´s unlikely you´ll use one of these in cooperation projects. Below is a quick start guide about the process which was used in Afghanistan.

رهنما یی مختصر بر ا ی اتصا ل پیپ ها ی پلاستیکی

تو سط ما شین بت فیو ژن ولدنگ

Quick start guide for HDPE
Butt fusion welding

سینتیا گو ارنا لیچ ـافغانستان ٢٠٠٣
Santiago Arnalich Afganistán 2003

جا بجا نمو د ن پیپ اولی

1.Put into position and secure the first pipe

جا بجا نمو د ن پیپ دومی

2.Put into position and secure the second pipe

بر يد ن انجا م ها ى پيپ ها

3.Cut off imperfections around the ends with the rotating cutter

امتحا ن نمو د ن انجا م ها ی پیپ

ها ی قطع شد ه

انجام ها ی پیپ ها با ید لشم بوده و بدون تیزی

و نا همو ا ری با شند

3b. Ensure that the ends are even, without burrs or rough edges

امتحان نمودن انطباق پیپ ها

انجامهای پیپ ها با ید به صورت دقیق

با هم برابر شده و خط مستقیمی را

تشکیل داده و خالیگاه بین شان اضافه

از ۱ملیمتر نباشد.

4. Check the alignment. The two pipes should meet in a straight line. The maximum gap should be 1mm.

انطباق غلط پیپ ها

شما ره ۱ انطباق غلط انجام های پیپ ها

را نشان میدهد طو ریکه پیپ ها با لا و

پا نین بوده بین شان پته پایه، تشکیل شده

شما ره ۲ شگاف اضافی را نشان میدهد.

Example of bad alignment which leaves pockets and ridges.

پاك نمودن انجامهای پیپ ها و صفحهء

حرارت دهنده (شماره ۵)

انجام پیپها و صفحهء حرارت دهنده باید توسط

پارچه،که در اتایل و یا پروپایل الکول تر شده با شد

پاك شوند. دیگر مواد پاك کننده با ید استعما ل نگردد

زیرا که پیپها را متضرر می سازند. بعد از پاکاری

به انجامهای پیپ ها دست نزنید.

5.Clean the ends of the pipes and
the welding plate (5b) with ethyl
alcohol. DO NOT USE OTHER
CLEANING PRODUCTS. They can
damage the pipe or prevent a good
join. Do not touch the ends of the pipe
once they have been cleaned.

5 Minutes ۵دقیقه

گرم ساختن صفحهء حرارت دهنده
درجه ۲۰۰ سانتیگرید را انتخاب نموده
برای ۵ دقیقه انتظار بکشید تا صفحهء
حرارت دهنده به حرارت مطلوب برسد.

Heating of the welding plate.
Select 200ºC, and let it heat
up for 5 minutes until it reaches
the required temperature.

2 Minutes ۲دقیقه

حرارت دادن انجامهای پیپها

صفحهء حرارت دهنده را بین انجام پیپها
گذا شته و اندکی فشار وارد کنید.

Heat the ends, placing the
heating plate between the
pipes and applying pressure.

۲ دقیقه 2 Minutes

وصل نمودن انجامهای پیپها

صفحهء حرارت دهنده را دور نموده

انجامهای پیپها ار بهم فشار دهید ,

طوریکه فشار تدریجاً اضافه گردد.

End joining. Remove the plate and join the pipes, gradually increasing the pressure at the joint.

15 Minutes

۱۵ دقیقه

جلو گیری از حرکت در وقت سرد شدن

فشار وارده را کاملاً رفع نموده پیپها را برای ۱۵ دقیقه شور ندهید. کوشش نکنید تا پیپها را

سرد سازید. شور دادن و سرد ساختن پیپها در این مدت با عث اتصال بسیار ضعیف پیپ ها

میگردد.

Immobilisation and cooling. Remove pressure and allow the pipes to rest for 15 minutes before moving. DO NOT ATTEMPT TO COOL THE PIPE. Unions that are cooled or moved too early are much weaker.

www.ingramcontent.com/pod-product-compliance
Lightning Source LLC
Chambersburg PA
CBHW070509200326
41519CB00013B/2760